Hope Is the Thing with Feathers

Hope Is the
Thing with Feathers

A PERSONAL CHRONICLE

of

VANISHED BIRDS

Christopher Cokinos

JEREMY P. TARCHER/PENGUIN

a member of Penguin Group (USA) Inc. • New York

JEREMY P. TARCHER/PENGUIN
Published by the Penguin Group
Penguin Group (USA) Inc., 375 Hudson Street, New York, New York 10014, USA • Penguin Group
(Canada), 90 Eglinton Avenue East, Suite 700, Toronto, Ontario M4P 2Y3, Canada (a division of
Pearson Canada Inc.) • Penguin Books Ltd, 80 Strand, London WC2R 0RL, England •
Penguin Ireland, 25 St Stephen's Green, Dublin 2, Ireland (a division of Penguin Books Ltd) •
Penguin Group (Australia), 250 Camberwell Road, Camberwell, Victoria 3124, Australia
(a division of Pearson Australia Group Pty Ltd) • Penguin Books India Pvt Ltd,
11 Community Centre, Panchsheel Park, New Delhi– 110 017, India • Penguin Group (NZ),
67 Apollo Drive, Rosedale, North Shore 0632, New Zealand (a division of Pearson New Zealand Ltd) •
Penguin Books (South Africa) (Pty) Ltd, 24 Sturdee Avenue, Rosebank,
Johannesburg 2196, South Africa

Penguin Books Ltd, Registered Offices: 80 Strand, London WC2R 0RL, England

Most Tarcher/Penguin books are available at special quantity discounts for bulk purchases for sales
promotions, premiums, fund-raising, and educational needs. Special books or book excerpts also can be
created to fit specific needs. For details, write Penguin Group (USA) Inc. Special Markets,
375 Hudson Street, New York, NY 10014.

ISBN 978-1-58542-722-2

Printed in the United States of America
1   3   5   7   9   10   8   6   4   2

Book design by Carla Bolte

While the author has made every effort to provide accurate telephone numbers and Internet addresses
at the time of publication, neither the publisher nor the author assumes any responsibility for errors,
or for changes that occur after publication. Further, the publisher does not have any control over and
does not assume any responsibility for author or third-party websites or their content.

FOR MY FAMILY

"Hope" is the thing with feathers—
That perches in the soul—
And sings the tune without the words—
And never stops—at all—

And sweetest—in the Gale—is heard—
And sore must be the storm—
That could abash the little Bird
That kept so many warm—

I've heard it in the chillest land—
And on the strangest Sea—
Yet, never, in Extremity,
It asked a crumb—of Me.

—Emily Dickinson, "Poem 254," ca. 1861

Contents

Introduction

ON AN AFTERNOON IN LATE SEPTEMBER, IN A BRISK PRAIRIE WIND, I WATCHED a bird I'd never seen before, a bird that had strayed far from its usual skies a continent away. Nearly epic in memory, that day began my journey, though I didn't know it then. The journey would take years and retrieve many things: first among them the name of the bird I had watched and didn't know—an escaped parrot that didn't "belong" in Kansas.

Seeing this bird led me to learn of—and revere—America's forgotten Carolina Parakeet, which once colored the sky "like an atmosphere of gems," as one pioneer wrote. The more I learned of the Carolina Parakeet's life, its extinction and its erasure from our memory, the more I wondered: How could we have lost and then forgotten so beautiful a bird? This book is, in part, an attempt to answer those questions and an effort to make certain that we never again forget this species nor the others of which I write.

For as I traveled to libraries and natural history museums on the trail of the vanished parakeet, I soon learned of other birds, other vanished lives: the Ivory-billed Woodpecker, the Heath Hen, the Passenger Pigeon, the Labrador Duck and the Great Auk.

I'd been a birder for a few years at that point in my life—the early 1990s—and I remain one today. Enthusiastic and, most of the time, at least moderately competent. (You can spot this kind of birder when he or she is in a crowd of experts; we're the ones who stay very quiet and will not venture a first identification for risk of mortifying embarrassment. "You thought *that* was an immature Black-throated Blue Warbler? Puh-

1

leeze!") Despite my quiet enthusiasm for birds, I knew nothing about *these* birds. And why should I have known? I couldn't see them, couldn't add them to my "life list." They wouldn't show up on the local Christmas Bird Counts. They were gone.

Their absences haunted me. I began to write. Nearly 10 years later, this book appears.

While most dedicated birders will pore over field guides to puzzle out the nuances of Little Brown Jobs (sparrows) or the maddening winter plumages of various gulls, I'd sit at a library carrel looking at John James Audubon's *Birds of America*. I would marvel at the improbably massive beak and almost pathetically tiny wings of the Great Auk or the spiffy black-and-white plumage of the mysterious Labrador Duck.

What had happened to these birds? And why? This book recounts the histories of six extinct North American birds, the stories of their lives and their demise, the stories of the people who killed the birds and those who tried to protect them—sometimes the same people. This is a personal chronicle, in which I weave together these accounts with my attempts to understand them—which led me to visit the places where these birds once lived and died. I take a journey into the past and emerge, sobered and saddened, but also fascinated. And resolved to grapple with hope in this environmentally complicated time: The birds taught me that we can learn from these losses, take comfort in what remains and redefine hope from "wish" to "work." We can work to protect the still-astonishing nonhuman lives that have come to depend on us for patience and care.

Not only did these extinct birds inspire me to delve into and present their natural histories, they also revealed many surprising, compelling *human* stories. There is, among others, the tale of the painter who saw the last confirmed Ivory-billed Woodpecker in the United States (as its forest was being logged) and the heretofore untold account of how the last known wild Passenger Pigeon died. These and other stories dramatize our various relationships to the land and the sky.

This subject matter also demanded engagement in cultural contexts, the broader historical forces that hastened these extinctions. Logging. The millinery trade. Unregulated hunting. Bird collecting by amateurs and professionals. The stories of the Passenger Pigeon and the Heath Hen illustrate, for example, the story of nineteenth-century urbanization in America, with the attendant development of voracious, concentrated public markets—markets that, for several years, sold unbelievable quantities of wild birds.

Perhaps unlike a professional historian and more like the poet I have been, I have found myself drawn to the oddments, the margins, so that a cookbook's reference to Passenger Pigeon Pie looms as importantly in this book as, say, logging statistics. A settler's account of how Carolina Parakeets in sycamores reminded him of Christmas trees in Germany— that matters to memory as much as facts of biology.

The book's concern with context continues to the present day, as scientists and others ponder the startling possibility that extinct creatures— including these birds—might be reborn through the rapidly advancing science of cloning. Such a prospect looms, of course, because of technological advance, real and potential. We consider this possibility, though, within a modern culture still struggling to discover some semblance of balance with the nonhuman world.

So I have tried to write not only a natural history, but something more—a chronicle at once personal and historical, a collection of factual narratives that engage where we stand now, in relation to the birds gone and the birds remaining. We may never restore vanished birds through the promise of cloning. That may remain a Hollywood fantasy. But we can restore—we can *restory*—these vanished birds to our consciousness. That can be an important act of recovery of the human spirit *in* the nonhuman world. I know I risk accusations of nostalgia (I do miss these birds) but anyone who sees the past as important territory—as a map for the present and the future—risks that allegation. Curiosity began my journey, which led to regret, which brings me always to wonder and dedication.

The Carolina Parakeet

Overleaf: Carolina Parakeets feeding on cocklebur, by John James Audubon.

The Forgotten Parakeet

ON A BRIGHT, CLEAR, WINDY DAY, MY WIFE AND I DROVE TO A NEARBY LAKE.
We recently had moved to the Flint Hills of eastern Kansas and, profoundly depressed, I yearned for southern Indiana, where we had been for several years, where we had become apprentices to the land, to still-beautiful fragments of hardwood forest. I disliked an office job I'd taken soon after we had arrived in Kansas, as Elizabeth began her teaching career at a state university. I missed the magazine editing I'd left behind. Also, I had a cold and felt, therefore, full of myself, literally and figuratively. A little anxious at my grimness, Elizabeth suspected that seeing birds would help. I blew my nose, dubious of nearly everything.

We thought we might see American White Pelicans, which look so ancient as they migrate through the eastern Kansas skies. That they occurred here was one pleasant surprise. Our destination for the day's brief foray was the River Pond, a small marsh adjacent to the reservoir in Riley County, Kansas, just outside the college town of Manhattan.

The blustery wind and bracing light soon helped me forget the lightness in my head, the heaviness, too, and the ache in my joints. Leaning against open car doors, Elizabeth and I watched an Osprey that had soared into view just as we were parking.

We stood by the banks of the Big Blue River. Here, the Blue serves as the reservoir's outflow channel. We watched cloud shadows plummet across the dam's earthen slope. A shaft of light touched the Osprey while I stood in a moment's gray. The sky kept changing. Talons dropping, the

Osprey would plunge to the water's surface in a flurry of spray, but we never saw the catch. This was the daily work of survival, but when the Osprey flew up suddenly on its crooked, sail-like wings—savagely distinct amid darting Double-crested Cormorants—it did so, I thought, simply because it could. That would be a supposition, perhaps even a romantic one, because I don't know the hunger that drives an Osprey to plunge and lift and soar and look and, turning, to try again.

Soon the ache in my soul vanished, too, at least for a time. The bright blade of the world cut painlessly and opened me. I would forget myself in what was about to happen and, over the years, this forgetting-of-self that comes from looking would lead me to love the bird life of Kansas, then Kansas itself.

We walked to the duck blind, a tidy shack on the edge of the River Pond marsh. We stepped carefully through a wet field, and I thought suddenly of Rachel Carson; her book *Silent Spring*, published in 1962, the year before I was born, alerted Americans to the poisoning effects of the pesticide DDT on songbirds, seabirds and raptors. By accumulating in fat tissues at ever-higher concentrations the further it went up the food chain, DDT had threatened with extinction some of the very birds I see today in Kansas. Peregrine Falcon. Osprey. *That* Osprey. I thought of Carson's struggle with cancer and her battle with chemical firms determined to slander her. She would not give in, and I felt minor, inconsequential, ashamed at the black hole of my self-pity.

All this flashed in my mind as quickly as a passing bird or two, a turn in the air, as I walked through a stand of plants I didn't notice then. Now I name and love Illinois bundleflower, whose seed pods cluster like tight brown fires.

It is tiring sometimes to think of our postmodern grief, but it is a crucial beginning, a necessary grief before the salve of some healing energy. One moment we are watching a life not our own—a raptor hunting, a firefly flashing, a bison grazing—and the next we cannot help but think of how we are the degraders, we the deciders. And, yes, sometimes, though not yet enough times, we can be the rescuers, the restorers.

At least for now. Until our species joins most of those that have burbled, walked, looped, soared, buzzed, spiraled, tunneled, whirred and floated on this 4.5-billion-year-old planet. The statistical paleontologist David Raup reminds us that 99.9 percent of all species that have ever lived on the Earth are extinct and that, on average, a plant or animal species vanishes after about 4 million years. A firefly's flash in the context of planetary history. "Life," writes Stephen Jay Gould, "is a copiously branching bush, continually pruned by the grim reaper of extinction, not a ladder of predictable progress. Most people may know this as a phrase to be uttered, but not as a concept brought into the deep interior of understanding."

The world keeps telling us the story of moments. That day, at the River Pond, blessed by the Osprey, uncertain of tomorrows, we watched the clouds congeal above the duck marsh. We scanned the skies. Behind the clouds the sun became a white circle, a pearl. The sun roiled in its thermonuclear fate, reactions so numerous we might call them infinite, though really they are only countless. Tenacious and ephemeral particles hurled to the surface and through the corona and through space in a kind of wispy premonition of the sun's own fate—a billion years hence—the vast swelling, as the sun becomes a red giant that will burn this Earth of life. We are hardly permanent. The bacteria in our guts are more permanent. Summer's pond scum is older by time spans we cannot fathom: The elder cyanobacteria are 3.5 billion years old. They are an exception to the reign of extinction because they are still here; they've beaten the odds. How long will they last, after we are gone?

More clouds came in, packing into available sky. A Turkey Vulture slid on the wind—a Great Plains front blasting through, maybe 30 miles per hour—and we shut ourselves in the duck blind beside the marsh. Out of the chilling wind, we were taken by its sudden absence from our skin, though the blind shook and creaked. As if our bodies are defined by what presses against them, self made self by what it is not. Our eyes adjusted and watered. We pulled out handkerchiefs and sat down on the one bench, unlatched the slats and let some wind back at us. I smiled at my wife, and she put her hand on my leg. We had never sat in this blind before—in fact, we had never sat in any blind before—so this newness seemed a small adventure.

Elizabeth and I looked out the peepholes across the marsh at bare trees. I felt myself in a vastness at once beautiful and unpredictable. I *felt* how the world is always this series of moments, this very one, extended into eras, epochs, beautiful names, those expanses *Cambrian Silurian Permian Triassic Eocene Miocene* . . .

And suddenly I saw this: a strange green bird I did not know. A color that was tropical and out of place in autumn Kansas, an exhilarating smear of green, bizarre to see it here! We puzzled a second, offering out loud that it must be a parrot or macaw of some kind. I imagined this black-and-green exotic flitting through an open cage door, an open window in some kitchen in Nebraska or Iowa, as it flew—perhaps it still would know—toward instinctive warmth. The strange green bird did not belong to this place or season, these cottonwoods. It was an escapee, a vagrant, a jungle's life.

We saw the bird zip and twist in front of us—it was black-faced, with what appeared to be black wing tips and a black tail, and fugues of a green, mostly dark green, it seemed, all over. The impression was quick, for a Sharp-shinned Hawk burst instantly from behind and above the blind. We heard the air closing behind its body as it hurled up toward the streaking green bird. The sky became a gray backdrop for pursuit. Across the pond's water a green reflection blurred.

We ran out of the blind when they flew from our sight. Seeing the sharpie—a big one, pigeon-sized—swoop up into a nearby tree, I turned quickly to find the mystery bird, seeing nothing, turned again, and there they were again: the hawk and the green bird, and just as suddenly, a *sec-*

ond green bird, the trio twisting and twisting in the sky, a live weaving of possible death, a braid. I lost them to distance, the hawk and its two prey.

Then, across the River Pond, a racket grew, a loud chitter, like a clattering of sticks, desperate and ceaseless, like the final ceremony meant to bring a lost god back. We scanned through our binoculars for vibrant green among the gray branches. There! Motionless but brilliant and noisy in the cottonwoods. We silently watched the green bird for several minutes. I noted the black face mask, the silvery beak, the dark green of the belly with (perhaps?) a lighter blue on the chest. Perhaps a trace of yellow on the underside by the legs? The back a vivid green. I heard the wispy rasp of what sounded like finches, and for long minutes the strange bird called, slowing down to tentative squawks, a kind of invitation to the other, absent one, and each squawk a beacon to the hawk. The desire to be paired again, to be *not* alone was, though, stronger than fear. The bird called outward, alone, to another of its kind, to learn how alone it was.

Until, sudden and poignant and tense, the second stranger—this bird whose green seemed from another world—came back, flew into sight, calling and calling, chased still by the persistent hawk, and the sky glimmered again with two green—parrots? macaws?—swooping sudden circles around a Sharp-shinned, who was simply hungry, desperate with its own need. This took seconds and then was over, there, in the world. The hawk, dizzied, broke off its chase. The strange green birds left us amazed beside an autumn wetland. Somehow they had escaped the hawk, somehow they had survived this hunt.

The next day another local birder confirmed my guess, but only after my first call to the Kansas Rare Bird Alert yielded no taped messages about weird, green birds darting around Manhattan, Kansas, of all places.

"Black-hooded Conures, that's right. Faster than greased snot," he said. "They've been around a few days. Probably wondering what the hell they're doing in Kansas. But probably glad to get out of a cage." I concurred, looked at the field guide and wondered aloud if these escapees would set up housekeeping at the River Pond. (They never did.)

Certainly I found something incongruous about colorful parrots flying over Kansas, a place—in my first years here—that I viewed through mental black-and-white celluloid. Parrots, conures, parakeets, macaws, all of them belonged in lush frondy places, jungles and tropics where the rich sipped umbrellaed drinks on stone terraces and where the poor sat on roadsides selling gaudy caged birds to tourists.

That, of course, was my ahistorical cliché. As I looked up descriptions of foreign parrots and parakeets that had escaped into skies they didn't originally belong to, I found references to a *native* bird just as green as the conures I saw. I read, for the first time, of the Carolina Parakeet—a *North American* parakeet whose green, yellow and reddish-orange plumage appeared vivacious and altogether quite wonderful. As stunning as I found the hawk-chased conures, *this* bird astounded me even more. That the Carolina Parakeet was extinct simply added to my amazement.

That I had never heard of such a bird did not surprise me; my recent interest in ornithology was eager and far from expert (and remains so). But others more experienced also did not know of the Carolina Parakeet. The more I spoke of the bird, the more it seemed that, somehow, its existence had been a chimera. Admittedly, my survey was small and unscientific, but intelligent people who could reel off the names of various dinosaurs and identify sparrows at epic distances could not name the forgotten parakeet. I realized, forcefully, what I suppose I knew abstractly: Histories, like species, can go extinct.

Later, I learned that our forgetting of the parakeet had begun even before the species was extinct. One researcher notes the story of a reporter who profiled an elderly Floridian recalling his life in the early 1900s. The old man remembered that parakeets would forage on his family's land and believed that the birds came from Cuba. That's what his neighbors had told him. But no—the birds were Florida natives; the birds were Carolina Parakeets.

The simple fact of the Carolina Parakeet—its name and its absence—stayed with me for several months, though I couldn't say much more about it, really. The Carolina Parakeet intrigued me, but its history had not yet fully captured my zeal. Details mattered less than my admittedly blurred image of the parakeet. I stayed busy with earning a paycheck and planning summer trips to escape Manhattan.

Several months later, wholly by chance, a tall, blond Finnish historian named Mikko Saikku came to town and delivered a talk at Kansas State University. His dissertation research concerned bird extinctions in the American Southeast, and he showed slides of dead Carolina Parakeets arrayed on trays. It had never occurred to me that I actually could *see* the bodies of these birds. For days I felt shaken by the sights from his slides: neat rows of colorful corpses, beautiful, inert. I remembered what the conservationist and writer Aldo Leopold once said: "Our ability to perceive quality in nature begins, as in art, with the pretty." I understood that the Carolina Parakeet—*Conuropsis carolinensis*—was, by virtue of its beauty and its extinction, more than just a species.

Slowly, those days, I had been learning to appreciate the loveliness of Kansas and of the Flint Hills, where the rare native tallgrass prairie still grows in wide vistas. My wife and I would be here a good while—I had come to peace with that—so I started paying more attention to the prairie and the birds it harbors, the Prairie Falcon and the Dickcissel, and even, at the lake in winter, many Bald Eagles. Perhaps Kansas was not as dull as my initial gloom had suggested.

So it occurred to me that Mikko Saikku's startling revelation—that Carolina Parakeets once lived in Kansas—might be a way not only to learn about the bird but to feel closer to the place in which I now lived. The Carolina Parakeet specimens he had photographed were held in a collection at the University of Kansas, only 90 minutes away. Perhaps someday I might be allowed to see those specimens. Before I'd go, though, I would do some research. Where, precisely, had the parakeet lived? How had it lived? And how had it vanished from the American sky?

As I began to learn the complexities of those answers, I could not have expected that a 10-year-old girl from another century would have mattered as much as she did. She joined me briefly, but significantly, in my search for the history of the forgotten parakeet or, rather, I joined her as I read of her journey.

For not far from the confluence of the Kaw and the Big Blue rivers, near where I had stood watching green conures outfly a hawk, young Sarah Dyer saw green birds, too. Along the then-expanding frontier, almost 140 years before I arrived at the River Pond marsh in a car, precocious Sarah had arrived with her family in this very area—where she saw flocks of Carolina Parakeets. Between her year and mine, between Sarah's autumn days in 1853 and my autumn day in 1990, it was likely that no other parrots had flown these local skies or graced these marshland trees. Between her day and mine, this continent changed.

Yet, by luck, we shared a place where each of us had sighted green birds above these rivers. The sense of connections I felt—to a vanished human being, to a vanished species, to the place we all shared, indeed, to time itself—branched upward in me like a cottonwood, which, with its furrowed bark and quavering leaves, I had come to cherish.

At 63, Sarah Dyer—married as Sarah Woodard—wrote down her memories of settling Kansas and therefore achieved a measure of unexpected immortality by her brief mention of what she called "wild parroquets." Ornithologists consider her record authentic, so it helped to establish the range of the only parakeet native to the eastern portion of the United States. This is what Sarah wrote:

East Spokane, Washington, March 8, 1906

My father, Samuel D. Dyer, was the first settler on the Big Blue. He came there in the spring of 1853, employed by the government to run the ferry at Juniata. In the fall the family moved there, traveling with teams from Ft. Scott, Kansas. We had one ox team and a team of horses. We drove our stock consisting of hogs, sheep and cattle. We could make but slow headway as the hogs and sheep could not travel very fast. We would camp nights and build a big fire and we children thought it great fun.

Game of all kinds was abundant. Buffalo—yes great herds of them. My brother-in-law, George Jameson, and Mr. Jacobs, killed a big buffalo on McIntire Creek. There was deer, wild turkey, prairie chicken, quails, wild geese, ducks and lots of wild parroquets when we first went there, but they soon left.

I could never again visit the River Pond marsh or drive near the Big Blue River without sensing green traces in the air.

In Sarah's days and in years before, the Carolina Parakeet thrived. Relatively common, it ranged across the eastern half of the United States, north to Illinois, Ohio, Indiana, Wisconsin and New York, down to the deep South and Gulf Coast states and west into Kansas, Nebraska and even eastern Colorado. The parakeet could venture that far west—across the prairies and plains—because it preferred what biologists call riparian woods, what writers of the nineteenth century termed more poetically "timbered streams." Such rivers and their trees snaked across the prairie and plains like narrow versions of the forests that were so vast back East. Where there were rivers and enough trees along the banks and in the floodplain, the parakeet could be found. Where there was river-bottom deciduous forest—including the cypress swamps of the South—the parakeet could live. In areas lacking wooded rivers and bottomland, the parakeet typically was absent.

Wherever it lived, it was noticed. Archaeologists once found in a prehistoric Indian mound an effigy pipe shaped like a Carolina Parakeet. For Native Americans, birds exuded spiritual importance. Creatures and forces of nature were gods to this continent's indigenous cultures; these things, including parakeets, manifested a sacred cosmos. Of course, the European settlers of the Atlantic coast and eastern forests (and the Spanish in the Southwest and Florida, for that matter) not only rejected such pantheism, they used its existence as proof of Christian and rational Enlightenment superiority. In turn, this justified genocide. The land, dangerous, seethed with savage pagans, snarling wolves, giant bears—yet it also revealed an inexhaustible plenty. The land was not sacred, but its riches were a heavenly miracle. And these arrivers in what was, to them, a New World often had eyes for wonder.

It is not surprising, then, that the Carolina Parakeet's noisy flocks and luminous plumage of green, yellow and red impressed writers of diaries, letters and books. A stunning bird, about a foot long from beak tip to tail tip, the Carolina Parakeet shone unlike anything in North American skies ever since. "We have seen no bird of the size, with plumage so brilliant; and they impart a singular magnificence to the forest prospect, as they are seen darting through the foliage, and among the white branches of the sycamore," observed a minister named Timothy Flint, writing about his travels in the Mississippi Valley during the first quarter of the nineteenth century. Parakeets dazzled Flint's children; the youngsters "contemplated with unsated curiosity the flocks of parroquets fluttering among the trees . . ." It's not hard to imagine why.

Consider how the Carolina Parakeet's chest and belly showed a green slightly deeper and more vivid than the green of an osage orange fruit or a black walnut's husk. Its body shimmered greens, like lake water or calm ocean in the cloud-shifting, low light of late day. Its back displayed a green like dark, shiny leaves. A portion of an even darker green wedged along the wing, and the wing tips carried hints of bluishness. Where wing met body—the coverts—a yellow streak sparked, another announcement of color in the air; the wing's outer edge tinged in an orangish-reddish tint.

Carolina Parakeets in the midwestern and western portion of the species's range were considered a subspecies or race, and these were paler than the more easterly birds. The western race had more pronounced yellow on the wings. East or west, the Carolina Parakeet's neck and head shone yellow, like the sawtooth sunflower, and its forehead shone the red of blood-oranges, the red of distant, slowly dying stars.

What bird of middle and northern Europe could compare? The Carolina Parakeet received its first extended published notice in William Strachey's *The Historie of Travell into Virginia Britania* in 1612: "Parakitoes I haie seene many in the Winter and knowne divers killed, yet be they a Fowle most swift of wing, their winges and Breasts are of a greenish colour with forked Tayles, their heades some Crmysen, some yellow, some orange-tawny, very beautyfull. . . ." Strachey concentrated on the bird's beauty but opened his description with an attribute of the Carolina Parakeet that settlers found nothing short of astonishing—the bird was strong enough to weather, well, the weather.

During the winter of 1832–1833, with temperatures in the single-digit range, the explorer Maximilian, prince of Wied, saw parakeets along the Wabash River in Indiana; the birds sought sycamores on which to feed. Parakeets near present-day Omaha braved temperatures of –22 degrees Fahrenheit. Elsewhere, they endured air as cold as –25. Just south of my home, in the walnut groves of Council Grove, Kansas, an Army unit encamped, in February 1847, in the midst of a snowstorm. The men received much cheer from the wild parakeets, who seemed to keep them company.

A few naturalists believed that the cold would sometimes induce the parakeets to hibernate, as bears do, but the evidence remains meager at best. Possibly, the birds became torpid, especially at night, in order to preserve precious calories. Certainly the parakeets showed themselves to be bright and noisy—cheerful, even—during long gray winter days.

Dumbfounded Dutch pioneers in New York beheld a flock of parakeets in the dead of winter—a sight that frightened rather than inspired. Women trembled and men covered their mouths. These shocking-green, feathered rockets shot about the forest near Albany in January 1780, a sign, a certain sign, of the end of the world.

Unlike the terrified Dutch, a German settler in eastern Missouri found the sight of winter parakeets reassuring and nostalgic. In a translation of his 1877 autobiography, Gert Goebel remembered:

> These flocks of paroquets were a real ornament to the trees stripped of their foliage in the winter. The sight was particularly attractive, when such a flock of several hundred had settled on a big sycamore, when the bright green color of the birds was in such marked contrast with the white bark of the trees, and when the sun shone brightly upon these inhabited tree tops, the many yellow heads looked like so many candles.
>
> This sight always reminded me vividly of a kind of Christmas tree, which was used [in Germany] by the poorer families . . . A few weeks before Christmas a young birch tree was set in a pail of water. In the warm room it soon began to produce delicate leaves. When on Christmas eve such a tree was decorated with gilded and silvered nuts and with apples and candies, it did not look unlike one of these bird-covered tree tops, only these enormous Christmas trees of the forest looked vastly more imposing than the little birch in the warm room.

So the parakeet was a comfort, at least to some.

From such reports, we can determine that Carolina Parakeets were so hale that they did not establish a regular migration from winter weather into more temperate regions. Apparently, the availability of food, whatever the season, most determined the movement of these birds.

Prior to the introduction of European-style agriculture, the Carolina Parakeet feasted exclusively on the natural abundance of a wild land. They would consume a variety of seeds, from cypress to pine to elm to maple. Nuts, such as pecans and various berries, including mulberries, attracted the green flocks. They nibbled on pawpaws, wild grapes and leaf buds.

Most decidedly, however, the species favored the common and widespread cocklebur. In what may be his most animated work, pioneering nineteenth-century ornithologist John James Audubon painted a family of Carolina Parakeets devouring cockleburs. One of the birds looks directly at the viewer, with his claw outreached toward us, as if we might hold the sustenance it needs. A reproduction of this painting hangs in my living room now. To fully appreciate the work, one must read Audubon's description of the procedure required for cocklebur dining:

> [The parakeet] alights upon it, plucks the bur from the stem with its bill, takes it from the latter with one foot, in which it turns it over until the joint is properly placed to meet the attacks of the bill, when it bursts it open, takes out the fruit, and allows the shell to drop.

Flocks of parakeets would reappear daily wherever they found cockleburs "until hardly any are left," Audubon explained.

The birds also relished salt, a necessary supplement to their diet. They would ingest the mineral at salt licks, springs and saline marshes. Naturalist Alexander Wilson—a failed Scottish poet who was Audubon's nineteenth-century contemporary and his competitor in compiling a multivolume work on American birds—once saw "great numbers" of Carolina Parakeets at Big Bone Lick, near the Kentucky River; there, he said, they drank "the salt water, of which they . . . are remarkably fond." The parakeets also consumed—as do other bird species—sand and gravel to aid in digestion.

To find seeds, fruits, nuts, salt and sand, the Carolina Parakeets would move about in flocks of a dozen to hundreds of birds. This roaming or "irruptive" quality makes it very difficult to estimate how many Carolina Parakeets once existed. While observers in or near the favored habitat of river bottomland forest found the parakeet common, even in such places the birds moved about as hunger dictated, much as Cedar Waxwings do.

If flock movements in search of food were relatively unpredictable, Carolina Parakeets nonetheless had order in their daily routine. Morning's light saw them chattering in treetops, as if tarrying; then they would suddenly swoop up and out to feed. The birds flew swiftly in undulant

Carolina Parakeets in treetops, from A History of North American Birds
(1874) by S. F. Baird, T. M. Brewer and R. Ridgway.

flocks until locating a suitable food source, all the while calling harshly
and loudly. Ornithologist Myron Swenk described the notes as a "shrill
series of rapidly uttered, discordant cries, given incessantly when the
birds were in flight, resembling *qui-qui, qui, qui, qui, qui-i-i-i,* with a
rising inflection on each *i* and the last cry drawn out." Perhaps when the
flock landed, a swamp or grove away, the birds then spoke to each other
with a gooselike cry, which, Swenk wrote, "was frequently uttered for
minutes at a time."

After such early-morning flying, feeding and conversing, the parakeets took refuge in trees and quietly loafed the afternoon away; while there, they muttered and chattered in low tones, doubtless finding reassurance in each other's talk. The parakeets preened their long tails and wiped their sharp bills against tree branches. Then, in the late afternoon, hunger would send them out to feed again before retiring for the night.

They spent their nights, like their days, together. Flocks roosted in trees made hollow from decay and disease. The parakeets grasped the inside of the hollow with their bills and feet, hanging in slumber through the night. Sometimes a lack of space meant that a few birds would have to sleep outside the cozy den; these birds clung to the tree, right beside the entrance hole. (We know nothing of the social "pecking order" of such flocks and whether the same birds had to sleep outside until a larger cavity was utilized or whether chance determined who slept inside or out.)

The Carolina Parakeet lived so gregariously that it also nested in colonies. No one ever actually saw a Carolina Parakeet nest, a fact I find amazing given how many naturalists wrote about the species. But the literature strongly suggests the bird used tree cavities for raising young. The parakeets may have lined these hollows with wood chips before laying their plain white eggs. Reports from Florida that the parakeets built stick nests in cypress trees can't be confirmed, but seem dubious.

There are other mysteries. What kind of courtship behavior did potential mates engage in? Did they nest in spring or summer? (The foremost authority on the species, biologist Daniel McKinley, infers from the contradictory observations of various naturalists that it was summer . . . but here, too, we'll never know for certain.) How many eggs constituted a clutch in the wild? What exactly was the breeding territory? How long was incubation? (McKinley suggests about three weeks, but admits he is working from incomplete sources. So sparse is precise information on the Carolina Parakeet's life that not even one wild egg of the species is beyond question: Eggs claimed as having been collected in the wild might have been from captive birds.) More questions abound: How strongly did mates bond? Was the rate of reproduction as low as records seem to indicate? Did individual parakeets really live up to 20 years in the wild, as it seems they may have? And on and on. The answers to such basic questions as these disappear into the forests like green birds among green leaves.

This seems an appropriate simile. Although boldly visible in winter-bare branches, although a flock could cover a salt lick like a wild green and red and yellow quilt, these parakeets all but vanished in trees leafed out for spring and summer. If the parakeets stayed quiet and still, they could hardly be seen in such deep green woods. In 1896, the naturalist C. J. Maynard wrote that the birds, so noisy in flight, "will all pitch, at once, into some tree and a sudden silence ensues. So great had been the din but a second before that the comparative stillness is quite bewildering, then too, the large flock of highly colored birds, lately so conspicuous, have disappeared completely."

Maynard, who was trying to hunt the birds, searched among the leaves for a clean shot, but grew short-tempered and exhausted in his futile looking. He simply could not find the flock. So he flung an oyster shell into the forest—and out burst a myriad of screeching parakeets. Their *qui-qui-qui* calls probably seemed like mocking. As the flock twisted and turned, as the individual bodies of the parakeets angled this way or that, Maynard could see more of them as they flew *away* than when the birds had been so invisibly close.

Only once have I seen a *flock* of wild parrots (parrots being larger versions of parakeets) and these I could not identify. Walking by a pier in San Francisco, near the parklike setting of Fort Mason, I heard a squawking I knew instantly was neither raven nor gull. I looked up to see—for just a second—a flock of perhaps a dozen parrots, with tails that seemed, at quick glance, rather long; the birds wheeled against the bright sky, then turned tightly into the green foliage of some large trees up on a far hill. Then, quiet. What were the birds? Where were they? I could not find them in the tree, though I scanned with a pair of small binoculars. I wished these new mystery birds would show themselves as the conures had done, years before, at River Pond.

Though Carolina Parakeets, of course, never ranged all the way to San Francisco, I could not help but think of them. When Carolina Parakeets wheeled into the trees, I thought, it must have been like this: all that garish flurry, then a vanishing. Unlike C. J. Maynard, I stood too far from the trees and anyhow had nothing to throw. These parrots would not scare, so, if they cared to, they could watch, quietly, the many humans below.

Exotic

Our Parakeets are very rapidly diminishing in number; and in some districts, where twenty-five years ago they were plentiful, scarcely any are now to be seen ... I should think that along the Mississippi there is not now half the number that existed fifteen years ago.

—John James Audubon, *The Birds of America*, 1840–1844

SOME 40 YEARS AFTER AUDUBON PUBLISHED THIS OBSERVATION, THE GREAT evolutionary theorist Alfred Russel Wallace would write, "The most effective agent in the extinction of species is the pressure of other species." The varieties of "pressure" that lead to an extinction are several, and to understand them we have first to gain a glancing familiarity with two terms: *ultimate cause* and *proximate cause*. Then we can see how they help to illuminate specific cases, such as Audubon's strangely disappearing parakeets.

For all the simple finality of an extinction—a species utterly gone—the complexities of reaching the moment when the last individual dies are myriad and fascinating. In any specific extinction, causation—Wallace's "pressure"—is like a complex unwinding rope that we have to weave back together, even if many strands are missing. (It probably would be useful to remember just what we mean by the term *species;* according to the American Ornithologists' Union, species are "genetically cohesive groups of populations that are reproductively isolated from other such groups.")

For many nineteenth-century ornithologists, the most important reasons for the Carolina Parakeet's decline in the wild remained essentially inexplicable. Others would have to tie up the loose ends of natural history more than a century later, finally giving us the sequence of events that threads through the life and death of this amazing species.

First, though, let's leave the parakeet behind and hear the story of the Hypothetical Wren—a species I've concocted to briefly illustrate the difference between ultimate and proximate causes. An apparently spunky bird, the flightless Hypothetical Wren lived only on a single volcanic island in the Pacific. When the volcano erupted, all vegetation died. Only four Hypothetical Wrens—three males and one female—remained.

The wrens had feasted on a species of spider that lived its entire life in the wide blossom of a single type of ground-loving flower. Now both flower and spider were gone, so the birds had to expend extra energy walking about and foraging, seeking seeds and insects that happened to blow in from other lands. (We're also assuming the volcano erupted just once, then went dormant. And geologists know that hypothetical lava cools and hardens almost instantly.) The new diet, however, was less nutritious than before, and the wrens weren't well adapted to foraging for unfamiliar food sources. They missed lots of easy pickings.

Furthermore, the aggressive males continually harried the weary female with their bullying courtship rites, which involved much flapping of stumpy, useless wings. Defying the odds, though, the lone female raised a brood in a rocky crevice. Soon enough a second female produced young of her own. It seemed as though the Hypothetical Wren might make it past this dangerous low point.

But the new diet and lack of genetic diversity severely reduced the birds' resistance to disease and cold, and so the population began to plummet. Wind patterns shifted, bringing far fewer seeds and bugs. Many wrens starved. Finally, just one lived alone amid stony desolation. Then a cyclone blew a cocktail swizzle stick off an oil tanker's bridge, hurtling the miniature spear at terrible speed directly toward the island. You can imagine the rest. The last Hypothetical Wren died. The species disappeared.

From this melodramatic tale, we can see that ultimate cause—in this case, the volcanic eruption—is that which first sets a population at risk for extinction by decreasing its size and/or its ability to reproduce. (Had the eruption destroyed the entire species at once we would have considered this a cataclysmic extinction event.) Had the eruption never occurred the later factors would not have come into play in a way that killed off the remaining birds. Ultimate causes can vary, of course, and sometimes aren't understood until well after the fact, if they're understood at all. Suppose that instead of a volcanic eruption a virus had attacked the male Hypothetical Wrens, shrinking their gonads to sterile nubs. That would have been ultimate cause, but scientists would have had to collect enough wrens to verify that this virus affected most or all of the male population.

Proximate causes are the later genetic, physiological and environmental disasters (such as errant swizzle sticks) that befall a species whose extinction already has been set in motion by the prime mover of some ultimate cause. Ecologists have ungainly names for various kinds of proximate causes, including one category called demographic stochasticity, which sounds rather like an adhesive used by census takers.

By bearing in mind our rudimentary distinction between ultimate and proximate causes, we can better understand the sequence of events in an extinction and determine the relative influences of various "pressures." In this way an extinction is not only a matter of natural history. It's a bit like a detective story, a murder mystery in which the victim happens to be an entire species.

In the case of the Carolina Parakeet, the most obvious suspect for ultimate cause is the loss of forest and swamp habitat. In a very literal way, American settlers could not see the forest for the trees because trees mattered as necessary commodities to a pioneering nation. Logging for wood fuel constituted the ultimate cause of the Carolina Parakeet's extinction, suggests environmental historian Mikko Saikku. He points out that wood-fuel harvesting surpassed lumber harvesting throughout much of the nineteenth century. The forests straddling the Mississippi and Ohio rivers, for example, "were central in the fuel-wood commerce. At the same time, the Carolina Parakeets of these areas were rapidly disappearing, while Florida—with almost no activity of this kind—still maintained its population . . ." Certainly clearing of forests for fuel, building supplies and access to waterways took a serious toll.

Other kinds of habitat destruction harmed the parakeet as well, including the conversion of swamps to rice plantations and the clearing of native bamboo grass for farms. Bamboo grew in thick canebrakes in parts of the Carolina Parakeet's range and provided food for this and other species. Far more important, the bamboo canebrakes might have triggered courtship and breeding among the parakeets, according to biologist Daniel McKinley. This species likely did not breed each year; the parakeets may have relied on some nonannual stimulus in the environment to induce reproductive behaviors. It's possible—though not proven—that when bamboo went to seed, the parakeets recognized this as a signal to find a hollow nesting tree and begin mating. (A species that has not been seen in many years, the Bachman's Warbler, also developed highly specialized behaviors to take advantage of the bamboo canebrakes.) With the bamboo stands cut down or burned off, the Carolina Parakeet may have lost an important cue that set successful reproduction in motion.

The clearing of habitats—primarily forests, but also swamps and canebrakes—visibly altered the American frontier. Yet few naturalists wrote of the connection between the destruction of habitat and the parakeets' decline. Ornithologist W. E. D. Scott's observation in 1888 that "the settlement of the country and clearing of land has served to disturb this species very much" barely registered among his peers.

Conducting extensive research on the Carolina Parakeet from the 1950s to the 1980s, Daniel McKinley recognized another, more insidious culprit. One of the mysteries of Audubon's comment about disappearing parakeets (he made the observation several years before it was published) was how early the decline seems to have set in—as early as the first third of the nineteenth century. McKinley wondered how it could be that some reports stressed the disappearance of the parakeets *prior* to the clearing or settlement of particular areas. Something terrible must have affected the species before the ax whacked into bark and heartwood, before smoke rose above bamboo thickets. Was there another force that began, almost invisibly, the downfall of the parakeet?

One of the clues that McKinley saw early on concerned the Carolina Parakeet's apparent adaptability. The bird moved hither and yon to find food, ate a tremendous range of nuts, fruits and seeds, and could tough out bad weather. In McKinley's words, the Carolina Parakeet had "evolved stability in a changeable environment." So whatever had caused the bird to decrease prior to logging out of habitat must have concerned a habit or need of the species that was not as flexible as, say, its diet. Biologists call this kind of very particular pressure or influence a "species-specific perturbation."

What had it been? Perhaps another bird species, such as the Passenger Pigeon, had outcompeted the Carolina Parakeet for food. Yet we know virtually nothing about how the parakeet interacted with, or ignored, other birds.

By the 1980s, when McKinley was synthesizing his life's work on the parakeet, a possibility he had raised years earlier seemed increasingly likely. What first had driven the Carolina Parakeet away from the hollow trees it needed for roosting and nesting was—of all things—an insect. A fast-spreading insect introduced by the pioneers: the European honeybee. The bee used the very same hollows that the parakeet did. "As a primary competitor for nesting and roosting sites, the honeybee barnstormed across forested eastern North America in colonial and early federal times, easily outdistancing the light-footed pioneers," McKinley writes. "Even settlers were sometimes surprised at the speed with which bees spread (most of them probably thought honey and wax as native a product of the continent as the Indian himself . . .)." In fact, Native American tribes could discern when pioneers approached because "the white man's flies"—European honeybees—arrived first.

To support the bee hypothesis, McKinley points to the Chimney Swift and the Purple Martin, two cavity-nesting species that abandoned hollow trees at virtually the same time the Carolina Parakeet began its decline. Swifts and martins could shift to hollowed gourds (early birdhouses), chimneys and other human structures. But the Carolina Parakeet, for some reason, could not adapt to new abodes. The parakeet apparently had evolved a singular reliance on hollow trees for roosting and nesting and could not alter that behavior. Unlike the Blue-winged Parrotlet, which, a continent south of us, fends off wasps from its nest, the Carolina Parakeet could not cope with a competing insect.

It may be impossible to say which mattered most in starting the extinction of the Carolina Parakeet: the cutting of the forest or the spread of the honeybee. If Saikku emphasizes the former and McKinley the latter, they each acknowledge the role of the other factor. The bees took over trees. Pioneers cut trees, for fuel and timber, for honey and beeswax. The Carolina Parakeets lost their homes because trees were felled or filled. The amount of available habitat lessened, and the chronicle of lost hollows began.

Unable to see how habitat loss and competition with the European honeybee initiated the decline of the parakeet, most naturalists of the nineteenth century and early twentieth century considered the ultimate cause of the bird's extinction to stem from shooting and live capture. We now know, however, that these very real pressures were proximate causes. They can be crystallized in three images: the flock, the gun and the cage.

Though equipped with formidable feet and beaks, Carolina Parakeets feared hawks and increased their odds of survival through flocking. But this genetic proclivity to togetherness—the adaptive need to be as one— proved to be an important factor in the bird's demise. When attacked by gunners, the parakeets would exhibit a behavior that only made the killing easier. The birds swarmed in disbelief.

"They flew around us in flocks, keeping a constant and loud screaming, as though they would chide us for invading their territory . . .," wrote the explorer John K. Townsend, noting an incident in Boonville, Missouri, on April 8, 1834. He continued:

> They seemed entirely unsuspicious of danger, and after being fired at only huddled closer together, as if to obtain protection from each other, and as their companions are falling around them, they curve down their necks and look at them fluttering upon the ground, as though perfectly at a loss to account for so unusual an occurrence . . .

Bullets generally work faster than behavioral adaptations; the parakeets could not help but fly over and about their dead and wounded kin, calling and looking, wheeling and screaming. While shooting parakeets, ornithologist Alexander Wilson remarked that the wounded birds would lure their companions back: "The whole flock swept repeatedly around their prostrate companions and again settled on a low tree, within twenty yards of the spot where I stood. At each successive discharge, though showers of them fell, yet the affection of the survivors seemed rather to increase . . ."

Though Townsend considered such slaughter "inglorious" and Wilson found himself "entirely disarmed" by the parakeets' behavior, many others did not. Anyone with a gun and a desire to shoot could kill scores of parakeets, an easy mark. Even those who had found the parakeet beautiful took aim. Shooting for the sport of it—even if it seemed unchallenging to some—served as a national recreational pastime, part of what one nineteenth-century writer called "a disgraceful greed for slaughter."

As pioneers settled down, cleared fields and sowed seeds, the bird adjusted easily to humans in one respect: The Carolina Parakeet broadened its palate by dining on crops. "They cling around the whole stack, pull out the straws, and destroy twice as much of the grain as would suffice to satisfy their hunger," wrote Audubon. One pioneer passing through Indiana noted as early as 1819 that "the parroquet commits depredations on the wheat in harvest, but it is a bird of uncommon beauty." It was reported that Carolina Parakeets often ignored the soft meat of fruits in order to get to the seeds; if they found the seeds were soft, Audubon said, the birds discarded the fruit altogether. All this contributed to a perception that the parakeets attacked crops in a kind of wasteful play. As the lives of families depended on the success of a year's plantings, farmers dealt with depredations by any creatures—from parakeets to wolves—at the end of a barrel.

Curiously, though, a number of sources one would expect to mention the Carolina Parakeet as an agricultural enemy fail to remark upon the species at all. While researching this book at the American Antiquarian Society, an important library of pre-1876 Americana, I found not one farmer's almanac or guidebook that singled out parakeets. Yet these are precisely the kinds of publications that should have highlighted the parakeets if the species had been a serious pest. At a time when the bird was common, Ohio Valley settler Daniel Drake doesn't have a word to say about crop depredations inflicted by the Carolina Parakeet, Daniel McKinley discovered, and Floridians writing of the citrus industry hardly refer to the species, even though some ornithologists claimed the birds ate oranges.

Although it's clear that the parakeets sometimes damaged crops, the extent of this activity is not at all certain. In any case, the bird made a very convenient target. "Even if the parakeet did not do much damage," McKinley observes, "it could be blamed for all of it." If they chose, farmers had little difficulty dispensing with dozens, even hundreds, of Carolina Parakeets flocking above the good, turned earth.

Probably the parakeets adapted locally to varying food supplies, perhaps eating corn on one farm, perhaps eating apple seeds on another. In some places they may have ignored crops altogether. On some farms, in fact, the parakeets actually may have been welcome. The naturalist Amos W. Butler claimed that farmers viewed the Carolina Parakeets with "high regard" because they ate the prickly cockleburs. Farmers found the cocklebur a difficult weed to control. With the parakeet gone, no other bird quite fulfills the role of consummate cocklebur destroyer; that is left to county noxious-weed departments.

Farmers who shot flocks of parakeets would either leave corpses to rot on the ground or find a use for them. Native Americans, settlers and city dwellers alike used all kinds of birds—from robins to Northern Flickers, from Cedar Waxwings to meadowlarks—as sources of feathers, folk medicines and food. The Iroquois, for example, incorporated the parakeet's feathers in belts, but Native American use of the species stayed within sustainable limits. Just what good a parakeet might have been medicinally isn't clear, though Alexander Wilson and others contended that cats died from eating parakeet innards; the liver may have contained concentrated toxins from cockleburs.

Certainly life in the wilderness and on the frontier demanded that people make use of whatever wild creatures were at hand, and city markets sold much of the same. Settlers ate just about anything if they had to, including parakeets, but some thought the bird much more than mere subsistence food. "A dozen of them make a most delicious sea-pie," wrote Christian Schultz in a passage from his 1810 travelogue. A German lawyer recorded in his diary of a Mississippi River trip that Carolina Parakeets "made a savory dish." Others thought the bird barely palatable, good only for fish bait.

People may have disagreed about the tastiness of the Carolina Parakeet, but few would dispute the beauty of its plumage. At a time when chic women considered feathers and even dead birds highly fashionable accessories, the parakeet was prized. We have no way of knowing how many Carolina Parakeets died to decorate women's hats and dresses, but the destruction of native bird life to that end was vast and indiscriminate. An 1886 *Science-Supplement* article conservatively estimated that some 5 million birds annually died to serve the fashion whims of "the 'dead-bird wearing gender.'" One scholar cites a 1902 statistic showing that a dealer shipped 1.5 tons of egret feathers from America to England to be used on hats. From these sobering figures we can glimpse the damage plume hunters inflicted. The millinery trade pushed several birds to the brink of extinction, including the elegant Snowy Egret, and plume hunters eventually even murdered wardens hired to protect bird colonies in the South.

On a visit to New York City, my wife and I saw a sign at the American Museum of Natural History that listed the feathers seen on women's hats in a single day on a single street in New York in 1885:

> Robin; Brown Thrasher; Bluebird; Baltimore Oriole; Blackpoll; Wilson's Warbler; Tree Swallow; Tree Sparrow; Bobolink; Meadowlark; Black-

Woman's hat with a dead Carolina Parakeet, dyed black, ca. 1890s.

burnian Warbler; Scarlet Tanager; Cedar Waxwing; Bohemian Waxwing; Northern Shrike; Pine Grosbeak; White-throated Sparrow; Scissortailed Flycatcher; and Pileated Woodpecker . . .

My transcription is partial. The arguments that raged in America and Britain over the millinery trade slaughtering so many birds helped shape the conservation movement in the early twentieth century. Bird clubs that agitated for state and federal bird protections often formed, at least in part, as a reaction against the plume hunters.

Two other kinds of "collecting" depleted the wild population of the Carolina Parakeet. Though it could not mimic human speech and did not sing lovely tunes, the parakeet was nonetheless a sought-after cage bird, especially among Europeans. And in the eighteenth, nineteenth and early twentieth centuries, naturalists and ornithologists also collected Carolina Parakeets, but preferred them dead. Just as all kinds of birds were shot as food and decoration, so, too, were they shot for display, the parakeet included.

In those days, the Euroamerican mind—at least of the "refined" class—valued collection of dead animals, *especially* those things believed to be doomed to extinction. Both for scientific descriptions and for social status, one had to have what was rare. Naturalists would justify the collection of eggs and dead specimens (usually called "skins") by appealing to the needs of science and education. Businesses that sold such objects thrived. Egg hunters even had their own specialized magazine, *The Oologist*. Displays of these objects—called cabinets—became points of private and civic pride as well as educational tools.

In the first issue of the short-lived but gorgeously produced 1830s magazine *The Cabinet of Natural History and American Rural Sports*, the editors wrote that "a more intimate knowledge" of natural history "will greatly increase the comfort and enjoyment of the whole human race." This belief is perhaps not so different from the contemporary emphasis on environmental education and experiences in the outdoors.

An appreciation for animals in their natural settings was part and parcel of the experience of the naturalist and sportsman, but the urge to collect, study and display the trophies, from hummingbirds to bison, typically went unquestioned. Not until the widespread use of field guides and binoculars, well into the twentieth century, would shooting birds for specimens mostly abate. It is worth noting that scientific collecting killed far fewer birds than did commercial hunting or other forms of shooting; in 1886, ornithologist J. A. Allen estimated that about a half-million birds—*of all kinds*—had been shot by ornithologists up to that point. The number pales in comparison to yearly total kills for the plume trade. In the 1920s and 1930s, American ornithologists shooting birds sometimes had to contend with shouts of disapproval from town residents and hard questions from the police. Attitudes had changed. (Shooting specimens continues today, on a much-diminished basis, because it helps scientists understand differences between species and subspecies and clarifies questions of bird distribution.)

Certainly the cumulative effects on Carolina Parakeets of egg collecting, shooting and capturing seemed clear to a few nineteenth-century observers. "There is little doubt but that their total extermination is only a matter of years," wrote three distinguished ornithologists in 1874. The extinction might even "be consummated within the lifetime of persons now living." Even as scientists roundly criticized sport and plume hunters, ornithologists continued collecting in order to obtain those valuable final parakeet specimens.

And the plume hunters, the sportsmen, the farmers? What might they have thought of the vanishing parakeet? The parakeets disappeared because they could not adapt to civilization. There was truth in that sentiment, though the nuances (and ironies) of that attitude escaped most. Some people thought that perhaps storms killed many birds or that flocks had migrated to South America. And if gunners had eliminated most parakeets, so much the better; a continent needed taming, after all. Regret had no practical value.

Habitat loss, competition with honeybees, shooting and capturing—these proximate factors lowered population size and survival rates for the Carolina Parakeet. Because it was a gregarious species, the Carolina Parakeet probably required a certain flock size to trigger courtship and mating. With flocks under siege, the birds had increasing difficulty locating appropriate habitat and initiating breeding. Without new birds entering the population, the species as a whole aged and lost genetic diversity. When a species contains a limited number of individuals, the loss of even one bird can shave years off the survival of the whole.

Often a population that breeds with an insufficient gene pool merely bides its time; limited genetic information can expose a population to disease by lowering resistance and introducing dangerous genetic mutations. There are in fact a few mentions in the literature on the Carolina Parakeet of birds suffering from some kind of apoplexy, whose cause is unknown. A limited gene pool also can lead to sterility. As populations shrink—as deaths exceed births—the vagaries of weather, from storm to flame, can wipe out many that remain.

Precisely when the Carolina Parakeet disappeared from the American sky depends on which report, which anecdote, which observation, one decides to accept. Though we know the outcome of the story—the parakeet died off in the wild, to survive only in cages—reports of scattered flocks and individuals, even as late as the 1930s, still tantalize. Perhaps these late, unconfirmed sightings appeal to the part of us that marvels at visions of cloning dinosaurs. We wish to deny, if momentarily, the finality of losses through some act of imagination. Like Tantalus, we reach for something that always recedes.

Missourian Gert Goebel noticed the absences early in the nineteenth century, just as Audubon had:

> Until the later thirties great flocks of paroquets came into our region every fall and frequently remained till the following spring . . . As the set-

tlements increased and the forests were more and more cleared away, these birds ceased to come. The few old settlers of the days, when the paroquets frequented these parts, feel just as little at home as those beautiful birds did; they long for peace and quiet, whether above the earth or beneath, it does not matter.

Extirpated from Kentucky, Tennessee, Alabama, Mississippi and Louisiana by 1880, the Carolina Parakeet was all but killed off in Arkansas, Missouri and Oklahoma by 1890. A few flew above the Arkansas River, then left, starved or were shot. The western race of the Carolina Parakeet diminished to single birds near Atchison, Kansas, in 1904, and another seen on the gate at the Notch, Missouri, Stone County Post Office on July 18, 1905. Postmaster Levi Merrill spotted the lonely parakeet. When he had lived in Oklahoma, he had seen flocks.

In 1912, two reports emerged from the interior of the species's range of parakeets that almost certainly had escaped from cages. One bird stayed for weeks near the Courtenay bottomlands, near Kansas City, then disappeared. The other parakeet was seen by a University of Chicago professor named Elliot R. Downing along the Lake Michigan sand dunes near Chicago. For nearly 15 minutes, on June 11, 1912, Downing watched the parakeet through his binoculars as it perched on a juneberry tree beside a pond.

My dictionary defines exotic as "foreign; not native," something that is "strange or different in a way that is striking or fascinating; strangely beautiful, enticing." In a way, America's Carolina Parakeet became an exotic in its own home range.

In Florida—and other scattered locales in the South—the Carolina Parakeet made, as we say, its last stand, though eventually the species became pitifully scarce as swamp draining and bird shooting accelerated. Done with a day's shooting, ornithologists and other collectors would lean their rifles against trees. The men wrote in cursive ink on small white tags the date, sex and location of the kill, then tied the tags with string to the legs of the dead birds. Because of these field records, we can attempt to know when and where the last wild-born, free-roaming parakeet died.

The 1983 edition of the American Ornithologists' Union (A.O.U.) *Check-List of North American Birds* reports this taking (shooting) of the Carolina Parakeet: The "last specimen taken in the wild [was] on the north fork of the Sebastian River, Brevard County, Florida, on 12 March 1913 . . ." According to Paul Hahn, who compiled an exhaustive list of the collected specimens of several kinds of extinct or nearly extinct birds, this specimen—a male—was shot by John F. Delaney. The dead parakeet now resides in the California Academy of Sciences in San Francisco, number 15714 in that collection.

But Daniel McKinley doesn't mention this date in an otherwise comprehensive report on the Carolina Parakeet in Florida. He believes that it may be "a specimen that someone sat on for several generations before admitting to the foul deed of collecting it . . . Maybe, of course, someone simply found data to go with a previously untagged specimen." The current chair of the American Ornithologists' Union *Check-list* committee, Richard C. Banks, isn't sure where other documentation for the March 12, 1913, date may exist. In fact, the 1998 edition of the *Check-list* drops this date and offers no speculation on when the last specimen was collected in the wild.

James Greenway's standard reference work, *Extinct and Vanishing Birds of the World*, further complicates the mystery of the shooting of the last wild parakeet. Greenway claimed it happened "on Padget [sic] Creek, Brevard County (east coast), Fla., [shot] by Dr. E.A. Mearns on April 18, 1901." (Four males were collected that day at Padgett Creek, according to specimen researcher Paul Hahn.)

Still other records from 1904 have been cited as the last taking of a Carolina Parakeet. But naturalist Errol Fuller wrote in his recent reference work, *Extinct Birds*, that the last known wild parakeet, a female, was shot on December 4, 1913, in Orlando, Florida.

The confusion doesn't end there. If the A.O.U. relied on Hahn's catalogue of extinct-bird specimens in claiming March 12, 1913, as *the* day, someone missed even *later* dates. Hahn noted, for example, a bird collected at Orlando that died on June 5, 1913, the skeleton of which resides at the Smithsonian. No collector or catalogue number is listed. Hahn also included a Carolina Parakeet of undetermined sex—feathers and all—in the collection of the Milwaukee Public Museum. Someone killed it at Osceola, Florida, apparently on October 20, 1913. It is number 3927 in that collection. And Hahn listed a female "skin"—a specimen, not a skeleton—in the Smithsonian collection taken at Orlando that died on December 14, 1913. No collector or catalogue number is listed. Perhaps this is the bird Fuller's text refers to.

There is, finally, in the matter-of-fact printed typescript of Hahn's work, on page 339 of *Where is that Vanished Bird?*, a skeleton of a Carolina Parakeet listed in the collection of the U.S. National Museum at the Smithsonian. No catalogue number. No sex determined. No collector named. This Carolina Parakeet was taken at Orlando, Florida, and it died on August 14, 1914. This must be it, then—the last wild Carolina Parakeet.

When I ask Daniel McKinley about the 1914 record, he suggests an intriguing solution to the mysteries of all these 1913 and 1914 specimens. "I checked out all of Paul Hahn's records," he tells me. The 1913 and 1914 records probably "came from one or another of [ornithologist Robert] Ridgway's flock that survived and finally ended up as specimens after many years in captivity." In other words, ornithologists or cagebird collectors had captured the birds alive in their Florida locales late in the nineteenth century; the birds died many years later in captivity—on the dates listed—and not in the wild. If that is so, they *weren't shot in the wild* on the dates listed in Hahn's book, whose format leaves room for confusion. The explanation eliminates all the 1913–1914 records from consideration.

So sometime in the 1901–1904 era, someone—we don't know who—

was responsible for the last shooting of wild Carolina Parakeets. Perhaps it was the respected ornithologist Frank Chapman. He saw 13 parakeets in April 1904 on the northeast side of Lake Okeechobee, in Florida, at a place called Taylor Creek. Of course, he killed a handful, even though he knew the species was nearly extinct. Perhaps fearful of condemnation, Chapman suppressed the fact of his 1904 parakeet killing in the 1914 edition of his *Handbook of Birds of Eastern North America.*

The finality of the very last specimen shot on a particular day ultimately eludes us. We can't know the date, the weather, the name of the collector.

Rumors and claims of sightings continued into the 1920s and 1930s. Plume hunters and scientists alike whispered about South Carolina and Florida, where some felt certain the parakeets persisted in backwater swamps, away from the advance of civilization. Many kept this feeling secret, for fear of other collectors or, in the case of ornithologists, fear of scorn. That did not deter ornithologist Harold Bailey from his bold claim in 1925 that he knew of "a small colony [in Florida] . . . but I do not expect it to last long, especially after this fact is published." One Henry Redding, who lived in Florida's Fort Drum Creek area, said he saw some 30 parakeets there in 1920.

One especially compelling report concerns the University of Florida Curator of Birds Charles Doe, who in 1926 claimed to have found three pairs of Carolina Parakeets in Okeechobee County, at Grapevine Hammock, in the wide Kissimmee Prairie. Doe didn't kill any of the adult birds. But he did take their eggs, which still exist. Further study, including genetic tests, could determine to what species these eggs belong.

Of these late sightings, the A.O.U.'s Committee on Bird Protection was skeptical. The Committee wrote in 1939, "The persistent rumors that Carolina Paroquets still survive in southern swamps have been investigated with some care by the National Association of Audubon Societies and others, with negative results."

Only three years before, ornithologist Alexander Sprunt Jr. and National Audubon Society colleague Robert Porter Allen thought they had seen a flight of Carolina Parakeets in the Santee Swamp in South Carolina. Their account is intriguing but ultimately unprovable, even though Sprunt later would get word that someone else had seen two adults and a *young* Carolina Parakeet in the vicinity. In the face of withering skepticism from other ornithologists, Allen eventually equivocated in this claim. Sprunt, however, remained convinced the birds they saw were not distant Mourning Doves but Carolina Parakeets. Regardless, a power plant development destroyed the swamp and any parakeets that may have lived there.

By the late 1930s, the culture of bird collecting had turned toward steep decline. The trade in wild bird feathers, meat and eggs—and random sporting slaughter—had virtually disappeared as public disapproval and legislative protection increased. Nature photography, like bird-watching, came of age. Books extolling these activities appeared as early as 1908, and Roger Tory Peterson's famous, handy field guide, first published in 1934, revolutionized birding. From the 1920s on, audiences across the country packed theaters for traveling lecturers who presented nature films and slide shows. Static cabinets of dead specimens yielded to field observation and pictures.

It seems fitting, then, that one of the last claimed sightings of Carolina Parakeets happened to be captured on film—purported to have been taken in Georgia's Okefenokee Swamp in 1937 by a guide named Orsen Stemville, at that time a resident of Crawford, Florida. When Stemville's rediscovered color film was presented in 1970 at a meeting of the American Ornithologists' Union, the audience watched carefully. If the birds were in fact Carolina Parakeets, then the species had somehow managed to live far beyond expectations. Accounts of the film confirm that the birds shown were parrots of *some* kind—but no one could say for certain that they were Carolina Parakeets.

Of course, the birds in the film might have been escaped cage birds, like the Black-hooded Conures I sighted at the River Pond in Kansas or the parrots I saw and could not identify in San Francisco. (I learned later that these were Cherry-headed Conures.) Such birds—escapees from illegal smugglers or pet owners—live as true exotics, birds that originate from Central and South America, Africa and Asia. We could say they don't belong here, but the exotic birds that have made new homes in the United States would disagree.

Monk Parakeets in Chicago's Hyde Park have built large nests and have attracted the support of neighbors and politicians whenever officials consider destroying the colony. Exotic parrots fly in Miami and St. Petersburg in skies once home to the forgotten parakeet. Cities far from the range of the Carolina Parakeet also host new exotics. If some consider them pests—no better than House Sparrows or European Starlings—others consider them friends. Red-crowned, Rose-ringed, Canary-winged, Blossom-headed, Yellow-headed, Orange-fronted, Green and Monk—all parakeets now in North American cities. Their bright colors flash across city skies.

Unlike the random releases mentioned above, the Thick-billed Parrot was deliberately reintroduced by wildlife biologists into Arizona. A native of the pine forests of Northern Mexico, the Thick-billed Parrot's range just touched Arizona. Release into that state of the Thick-billed from captive birds and stock captured from smugglers seems so far to be a success, but the future remains uncertain. Officials have worried about the well-being of some birds—those that had been stuffed behind hubcaps by smugglers driving toward pet stores in the United States. Perhaps conservationists will be able to use the Thick-billed Parrot's reintroduction to help spur even stricter controls against the voracious illegal parrot trade, which is "pound for pound" more lucrative than cocaine trafficking, according to one U.S. Customs official.

This story and others from the Caribbean demonstrate that activists, scientists and local residents—working together—can promote sustainable ecotourism projects in native parrot habitats across the world, which can help those in desperate poverty elevate their living standards. After all, worldwide, 25 species of macaw, parakeet and parrot might join the forgotten parakeet, if such projects are not developed.

I would like to see Thick-billed Parrots in the wild, to see a native parrot above the land in which it first evolved. For now, I must be content with having watched three Thick-billeds at the Cincinnati Zoo, where the birds roost on a fake indoor cliff. As I gazed, two preened on a ledge and clasped bills while the third slowly climbed up a narrow crevice using hooked bill and claws. Their green plumage accentuated the red on their heads and their big white eyes. The birds shrieked and eyed me warily. The climbing parrot suddenly plummeted to the ground—all afluster and afeather—then began sluggishly to climb again. I recalled descriptions of the slow pace of Carolina Parakeets sidling along branches.

One of the parrots slowly opened a claw—as if grasping for air, as if aiming at me—then picked at the claw with its bill. For a moment—and I shook my head in disbelief—it was as if John James Audubon's painting of Carolina Parakeets, the one on my wall at home, had come to life before me.

Hope in a Cage

HERE GLEAM THE EYES OF A LIVING CAROLINA PARAKEET. THE BIRD'S HEAD pulls back, slightly, looking at us from aside, in profile. The pupil—a large dark spot, a black hole—gathers light in the moment and sees. The parakeet perches in a tangle of cocklebur, its relished food. That could be sky behind the bird, haze or clouds, some milky light, white as salt. This could be a wild bird beside a marsh in Kansas or at the edge of a farmer's field in Georgia.

The white behind its body is, in fact, a backdrop, a prop or wall, though the bird was no specimen, no mounted corpse. This bird, this Carolina Parakeet, breathed and, in its way, thought before, as and after a shutter opened to capture the image. Naturalist and physician R. W. Shufeldt posed the picture at the turn of the century, and I first saw the photograph reprinted in T. Gilbert Pearson's mammoth and delightful *Birds of America*—the first time I'd seen the image of an actual, living Carolina Parakeet. I sat with the book at a cramped desk in a library and stared for a long time.

The story: Shufeldt had borrowed a pair of birds, a male and female, from one Edward T. Schmid. Dr. Shufeldt conveyed the agitated birds to his home and, after several difficult hours, managed to get one just calm enough that he could snap the photograph. It is a beautifully haunting and vigorous composition, redolent of rivers, air and flocks.

The two birds later died when they chewed on the cage that Schmid had just painted for them. He had the birds stuffed, and I do not know the fate of the bodies.

Live Carolina Parakeet perched in cocklebur, in 1900.

Surely that day, as Shufeldt tried to lull the parakeets, he thought back to just a few years before when he found himself on a train stopped at mid-afternoon, in 1884, somewhere in eastern Kansas. Dr. Shufeldt gazed at bountiful corn gathered after harvest just 10 yards from his window and stared in disbelief. "Much to my surprise, there was perched upon this shuck a fine specimen of the Carolina Paroquet," he wrote of the incident. The parakeet—hungry and alone—attacked an ear of corn dried by the summer heat.

As the species disappeared from places in which it had lived for thousands of years, individual parakeets unwittingly found themselves in new homes—the homes of people like Edward Schmid and other aviculturists for whom the keeping of captive birds was an interesting and gentlemanly avocation, a part of the natural-history collecting craze.

The Carolina Parakeet proved to be a relatively easy species to keep in cages. Wrote ornithologist Thomas Nuttall in 1840: "The Carolina Parrot is readily tamed, and early shows an attachment to those around who bestow any attention on its wants; it soon learns to recollect its name, and to answer and come when called on . . . As a domestic, [it] is very peaceable and rather taciturn." Owners found that these attributes compensated for the Carolina Parakeet's inability to mimic or sing. Some of these owners may have followed Audubon's advice on how to calm unruly, newly captured parakeets: plunge the birds repeatedly in water.

Although tamed, the parakeets seemingly could not quite forget their wild origins. Through anecdotes and stories, we can witness some of the behavioral adjustments to captivity triggered in—or perhaps worked out by—the animal brain and emotions of the Carolina Parakeet. Thomas Nuttall provided this example:

> One which I saw at Tuscaloosa, a week after being disabled in the wing, seemed perfectly reconciled to its domestic condition; and as the weather was rather cold, it remained the greater part of the time in the house, climbing up the sides of the wire fender to enjoy the warmth of the fire. I was informed, that when first caught it scaled the side of the room, at night, and roosted in a hanging posture by the bill and claws; but finding the labor difficult and fruitless, having no companion near which to nestle, it soon submitted to pass the night on the back of a chair.

Domesticated Carolina Parakeets often had companions of their kind because aviculturists preferred to own a pair of parakeets (though often the sex did not matter to the owners). Nuttall wrote that parakeets in outdoor cages attracted the interests of their wild kin, who pressed against the bars to snuggle against their unfortunate friends, "even thrusting their heads, at such times, into the plumage of each other."

Such sociability suggests that captive Carolina Parakeets not only were easily kept but also easily bred. If we now see captive breeding as a moral and ecological imperative for endangered species, aviculturists of the nineteenth century found parakeet breeding little more than a part of their hobby.

The report of a Dr. Nowotny illustrates this lack of seriousness. In an 1898 issue of the respected ornithological journal *The Auk*, Dr. Nowotny related that his pair of Carolina Parakeets produced 10 eggs, 7 of which were destroyed either by the birds (inexperience) or Nowotny's family ("carelessness"). There seems to have been little interest in assisting the inattentive parents with the job of incubating eggs. The three that hatched all died. These failures hardly distressed the doctor, who preferred instead to focus on the birds' tameness:

> The old birds are very devoted to each other and are always together, and if one flies away the other follows immediately. They stand cold very well, but enjoy having their under parts touched by warm breath, for which purpose they cling to the wires and permit me to breath [*sic*] upon them, pecking me on the nose tenderly at the same time. In the cage I can play with them as I wish and even take them in my hands, but I dare not grasp or close the hand, for then they slip away at once, screaming.

Because the birds loved music, Dr. Nowotny would play songs to calm or distract them.

Although the Carolina Parakeet would breed in captivity, owners exhibited a startling lack of rigor for essential concerns, such as determining the best diet for the bird or for creating social conditions conducive to reproduction. This lack of rigor doomed efforts to propagate the species in any significant way.

Furthermore, no one could yet see the larger context in which captive breeding could take place: through a widespread, well-managed program in which private owners and zoos shared birds for breeding. Such a program might have prevented the extinction of the Carolina Parakeet—the only species in its genus, *Conuropsis*, which is therefore also extinct. Instead, aviculturists and zoos competed against each other for prized individuals, knowing that the species kept inching closer to extinction.

"*They* had their chance . . .," writes an angry Daniel McKinley. "Their records show a series of disappointments and a heartbreaking waste of eggs and of young birds and old, the loss made even more serious by a failure to keep adequate records of just what did go on."

Edward Maruska, the director of the Cincinnati Zoo—which once owned Carolina Parakeets—believes the species not only could have been kept alive in captivity but perhaps even released again into the wild.

Amazingly, one of the ornithologists who predicted the extinction of this species—Robert Ridgway—himself owned domesticated Carolina Parakeets, which mated and hatched out young. Yet not even an eminent scientist like Ridgway built on this success. He never developed contacts with other Carolina Parakeet owners, who might have created a modestly diverse captive-breeding flock.

Ridgway's parlor-room brood sat in a basket lined with white cheesecloth. The strange, ugly creatures—four of them—looked more like baby dinosaurs than anything else, their skin wrinkled and mottled, their heads thick and big. Ridgway insisted to Paul Bartsch that he take one of the Carolina Parakeets home for himself as Ridgway could not possibly care for all the nestlings, and, besides, the parents continued to ignore this one baby. Had Bartsch not accepted the bird as a pet, whom he named Doodles, it would have died.

A pioneer in bird banding—and the translator of Dr. Nowotny's article—Bartsch would later write of Doodles with the kind of imagery and tone reserved for a loved member of the family. He writes of his pet parakeet flying outside, temporarily escaping, chasing after pigeons for, he thought, some company; of Doodles crawling beneath his blanket, cuddling his neck, sleeping with him; of Doodles snuggling a pet squirrel in a box near the furnace; of the bird trying to steal baubles from a dresser top. Bartsch wrote of Doodles falling from his door perch, dying, looking at his owner who held the bird "gently [as he] slowly closed his eyes and stiffened . . ." Bartsch donated the body to the Smithsonian's extensive collection of Carolina Parakeets, probably in August 1914.

One autumn afternoon, when I opened the mail at my desk, I found among the utility bills, magazines and solicitations from environmental groups a thick package containing photographs. Needing research queries answered, I had tracked down Daniel McKinley, who had filed away his stacks of parakeet research notes after his retirement from the State University of New York at Albany. In among the files were photos of Ridgway's flock, taken in 1902, by Paul Bartsch. I had never seen them. Only one picture had been published and no library likely to have the original glass-plate negatives could tell me that they still existed. Now McKinley was graciously sharing the prints with me. So I spread the photographs on the desk and gaped. I felt like "some watcher of the skies," though what swam into my ken was a single remnant of something wildly glorious, a species—and not one photo showed a sky, though bright sun filtered through the unseen windows.

Here, Doodles perches on the handle of a basket filled with nuts, a nutcracker and two fruits—pomegranates, I think—with the basket itself placed on a wicker chair. In another scene, the parakeet sits atop the handle—it looks like silver made to resemble twigs—and brings up a foot for preening while the other grips its perch. In several of the photos, Doodles is blurry, having moved during the exposure, a photographic flaw giving evidence of life.

In the last of Bartsch's pictures, taken in 1906, Doodles has clambered to the face of a "Mr. Bryan," holding on to his white dress shirt and polkadot dress tie. Some of what little hair that is left on Mr. Bryan's head has been mussed up, probably by the playful, daring Doodles. Though Doo-

*A "Mr. Bryan" with Doodles, the pet Carolina Parakeet, in 1906,
photographed by Paul Bartsch.*

dles's head hides the gentleman's lips, it appears that beneath his mustache, Mr. Bryan might be kissing the bird, this beloved parakeet, on some sunny American afternoon.

The final years of the life of this species comprise a tale of wild uncertainties and domesticated diminishments. The only irrefutable end we can cite came when the last known living Carolina Parakeet—named Incas—died on February 21, 1918, at the Cincinnati Zoo.

Perhaps on the night of February 20, a few wild parakeets—stragglers, exotics—slept in the warmth of an unknown swamp tree in the deep

South. Perhaps some pet Carolina Parakeet, unknown to history or science, cuddled the bars of its cage and lived for days, weeks or months longer.

Just two days before, the city had enjoyed, despite a half-inch of rain, a balmy 62 degrees, a fine day for February, a taste of spring, a day of warming cells, of mud, of emergent buds. Then sometime on February 20, a cold front moved through, with snow spitting in the air. The temperature dropped to 7 degrees. The harsh cold lingered the day of Incas's death, though it would warm above freezing the day after. By month's end, it was over 70 and spring seemed to have finally arrived.

One of so many virtually unnoticed documents of American bird extinctions, an article from the Friday, February 22, 1918, *Cincinnati Times-Star* gives us a keen sense of how the zookeepers felt the loss of Incas. The story's headline read, "Far-Famed Last Parrakeet of its Kind is Mourned at Zoo." The subhead read, in part, "Grief Was a Contributing Cause."

> A student of bird-life, acting as coroner in the case of "Incas," the Carolina parrakeet, said to be the last of its race, might enter a verdict of "died of old age." But General Manager Sol A. Stephan of the Zoo, whose study of birds goes farther [*sic*] than mere physical structure, development and decay knows the bird died of grief. "Incas," coveted by many zoological gardens, died Thursday night surrounded by his genuinely sorrowing friends, Col. Stephan and the keepers. Late last summer, "Lady Jane," the mate of Incas for 32 years, passed away, and after that the ancient survivor was a listless and mournful figure, indeed . . .

And with the death of the last Carolina Parakeet—whether it was Incas or some wild bird in Florida or South Carolina—some half-dozen feather mite species that lived only on the parakeet probably also went extinct. ("An entire ecosystem," biologist E. O. Wilson writes, "can exist in the plumage of a bird.")

We know relatively little of the life of Incas and Lady Jane. No photograph of either one exists. The Cincinnati Zoo paid $40 in the late nineteenth century for 16 Carolina Parakeets caught in Florida; Incas and Lady Jane were thus valued at $2.50 per bird. Later, the London Zoo offered $400 for the pair.

We know there were reports that Incas and Lady Jane would throw their eggs out of the cage's nest. I never found, however, any indication that the zookeepers tried to incubate those eggs, hatch them and raise the young. Perhaps they did. Perhaps the keepers spread cushions beneath the hollow tree they gave to Incas and Lady Jane, a tree in which the birds would pass the nights and, apparently, for a time, mate. From the cushions the keepers might have gathered eggs.

For a long time, Incas had been so utterly forgotten that authorities misreported (and sometimes still do) his death date as September 1, 1914—the day that Martha the last Passenger Pigeon died, also at the Cincinnati Zoo. After an exhaustive search, *Audubon* magazine writer George Laycock and Ohio wildlife artist John Ruthven determined the correct date of Incas's death as February 21, 1918.

They also had hoped to find Incas himself, whose body had been promised to the Smithsonian but never arrived. Incas, according to Laycock and Ruthven, may be an untagged specimen at the Cincinnati Museum of Natural History.

For me, the story of the Carolina Parakeet does not end with the end of the species. For months, I've been anticipating a trip to a museum where I'll be given access to a specimen collection off-limits to the general public. Touching a Carolina Parakeet is, I've been telling myself, the best way to make this loss actual.

My wife and I have stopped at the Museum of Natural History, at the University of Kansas, on the drive home from the airport in Kansas City. Yesterday I was in Indianapolis, visiting my father in order to begin a delicate reconciliation, a retrieval from the silence that has gaped between us. He's retired now and, together with his wife, drives a motor home around the country. They look at birds. We could talk about that. It seems this trip is a history of loss, personal and continental, but on this flight of museum stairs, I'm also thinking how tired I am. And now I will face the birds that have compelled me for so long. Just downstairs, among the exhibits, I saw part of a hollow cypress tree that once sheltered parakeets.

I walk down a narrow aisle cluttered with cabinets and boxes. A thirty-something man with glasses, dark hair and a mustache chats on the phone in his office. Curator and ornithologist Town Peterson gets up, smiles, shakes my hand and small-talks pleasantly. Having finished grading for the semester, he's writing up a history of ornithology in Mexico.

"I want to show you something," he says, smiling. We walk through tiny pathways between tables and cabinets—the place looks like a high-school science room that's been used as a fallout shelter. Peterson pulls out a small, clear bag from a cooler. The bag contains a tiny brown feather. "Know what this is?" he asks, and I shake my head. I'm thinking he could not only identify the species, but age and sex the bird, from this one small sample.

"It's from a moa."

There were once 20 species of moa—all rubbed out by the Maoris, a reminder that human-driven extinction events occur among a variety of cultures. Scientist David W. Steadman has shown that prehistoric islanders in the Pacific killed off some 2,000 bird species, diminishing by one-fifth the global number through a variety of activities, including habitat destruction. The differences today are in the scale of our extinctions, our consciousness of them and our developing understanding that time is the deepest wilderness in which we wander.

"But you came to see parakeets."

Dr. Peterson walks to a cabinet and pulls out a drawer. I glimpse a dozen or so Carolina Parakeets. What am I supposed to feel? Do I prepare myself to feel something? Is this now just too easy: See the birds and feel poignant? What kind of epiphany would that be?

Peterson walks through the maze of the room with the drawer balancing on his head, one hand holding the drawer and tilting it when he has to turn a corner. The parakeets slide. I imagine them tumbling to the floor, hoping that such an ignominy doesn't occur. He sets the drawer on a table and excuses himself to talk with a student who's come to complain about her grade.

I hesitate a long time, not sure if in picking up a Carolina Parakeet I will cross some threshold I can't identify or name. I stare. "They must have been beautiful to see," Peterson says, arriving again behind me. I nod. The puffy gray marks I see on the bodies are shot marks. He picks up a bird with his bare hands. "See how this one has both new feathers and some old, tattered ones?" Again I nod. "This could be an older juvenile bird, molting in April or May." I ask him why parrots have so few feathers, in comparison to other birds, and he answers: "Not everything has a reason for being." He excuses himself again, and Elizabeth and I stand quietly for a very long time.

My hands do not, do, want to touch them. I lift a long tag knotted to a parakeet leg. A female, shot by N. S. Goss on October 21, 1875, in "Indian Nation"—Oklahoma. Another sudden connection—Goss wrote the 1891 *History of the Birds of Kansas*, which sits on my desk at home.

My lifting the tag causes the bird's body to shift a little, a movement, and I imagine all 16 of these "specimens" as living birds. A flock in a winter sycamore. No, in a bald cypress. Most of the birds were shot on March 8 and 12, 1893, in Florida. Perhaps it rained from March 9 to 11. Most of the specimens are female. I wonder if there were any nestlings that died after the females were killed. Probably not. March would have been quite early for laying and hatching eggs, which likely occurred in the summer.

This is startling: how their heads are hollow—I can see through a breezy skull. (Some are stuffed with cotton.) All their heads hang limp in death, necks bowed, as if in grief. I think of the flock: the first one shot down, the others screaming, curving back, flocking toward that wounded one, screaming.

I lift one up; suddenly, I lift a parakeet. It is *so light*. The greens still bright and variegated, the barbs and barbules defined, precise—imagine the preening after morning feeding, the wings outspread, folded in, the head cocked to see. I can look in the hole where the left eye once blinked.

I am caught in a matrix of awe, grief, disgust and desire. I put this specimen down, pick others up, begin taking pictures. Those beautiful green bodies against the white tray seem somehow obscene. Then I notice a tiny bit of feather has rubbed off a parakeet, perhaps from my picking the birds up, putting them down (we get used to anything), and I reach my slightly sweaty forefinger out to touch and to retrieve that feathery dust at the bottom of the wide, white drawer. I could take this, I think, I could take this home. I could own this, encase it in lucite, place it on my desk where I keep round, smooth stones from the Bay of Fundy, a chunk of rhyolite from New Mexico's extinct Capulin Volcano and a polished black trilobite. My throat feels dry and scratchy from airliner air and from the realization that now I understand a collector's mentality. What harm would it be to have this tiny feather from this dead bright bird? None? Maybe. If I kept the parakeet feather, I could talk about it at dinner parties, show it to friends and visitors. I want very much to have it, but I do not say a word of this to Elizabeth for fear that she might say, yes, it would be okay, or, no, how could you think such a thing?

I put the feather down.

Their beige mandibles, their crowns kissed with orange, their yellow heads and necks, a yellow like the yellow of some roses, and their bodies of such various greens. Even in death the parakeets have a stunning beauty, even the grayish brown of the underside of the tails. I write down descriptions, details, look at the birds, then close my notebook, finished.

Peterson gives me two chapters from his book on ornithology in Mexico. I take them gratefully, thank him for the visit and watch him slide the

16 Carolina Parakeets into a cabinet as dark inside as a hollow cypress at night. As we walk across the floor, I hear the sounds of humans in buildings on a Friday: the squeak of soles on tiles, the hum of lights and computers, the friendly banter of colleagues, the whine of a door hinge that needs oil. My wife and I walk to our car beneath a blue sky, across which starlings fly.

———

In the exceptionally hot and dry August of 1904, politics, burglaries and the Japanese war with Russia were big news in Atchison County, Kansas. On one of those August days, on his fruit farm near Atchison, even closer to a town called Potter, Wirt Remsburg aimed his gun at a lone, chattering, bright-green bird—a Carolina Parakeet—and fired a shot that killed one creature and brought a species that much closer to the stuffed oblivion of lit museum cases.

Wanting to see where it had happened, I drove to Atchison and Potter. I wondered who, if anyone, remembered.

Potter was a town of ramshackle homes and trailers, a parcel of Ozarkian poverty. It offered nothing. "Let's just keep driving," my wife said, and I agreed. I saw no signs for a Remsburg farm. What was I expecting anyway? An apple tree with a wooden arrow pointing to the precise branch from which the bird fell? We drove along the hilly roads of lovely northeastern Kansas, back to Atchison, to the local library and historical society.

Wirt Remsburg's brother, George, a journalist, archaeologist, collector and naturalist, wrote a letter to the editor of a publication called the *American Society of Curio Collectors Bulletin.* The editors summarized a few details of the 1904 incident and published them in a brief account, dated May 15, 1906. ". . . [T]he bird having alighted in the thick branches of an apple tree . . . was mistaken for a chicken hawk," the article noted. George, the editor added, "much regretted the fact that the specimen killed by his brother was so badly torn to pieces by the shot that it could not be mounted." The lone bird had chattered so loudly and for so many days that many people had noticed it.

The editor opined that the disappearance of the species was a mystery. From the many reports of the abundance of the Carolina Parakeet in the bottomlands of the Missouri River, reports throughout the nineteenth century, it had come to this. One bird.

Walking by the blocks of grand and genteel Victorian homes—built by river money, by rail money—I watched the first leaves of autumn falling onto Atchison's red-brick streets. No one at the library or the historical society knew about the shooting, let alone the species itself.

How often do we memorialize these final acts, these last blows in our murder of a species? There are a few memorials to the extinct Passenger Pigeon, including one at the Cincinnati Zoo, which also honors the Carolina Parakeet. I know of no other monuments to this green bird. One can buy, though, a garden bench, from a South Carolina company, that includes in the elegant sweep of its metal arms, a replica of a feeding Carolina Parakeet. The design dates to the nineteenth century.

How much more surprised might we be, how much more engaged, if we found among fields and trees and neighborhoods a memorial to Remsburg's killing of one of the last western Carolina Parakeets? Maybe some who would read such a sign could realize, even a little, that extinction is not some newsbrief from a distant rain forest. It's here. It happened *here*.

In Atchison and Potter no such plaques exist. No interpretative trails, no self-guided walking tours you take with brochure in hand. No centers devoted to explaining the life of the Carolina Parakeet. Without these things, however, or some version of them, the forgetting becomes nearly final at the very place where the act occurred, the very place where memory can be a local guide and personal catalyst.

Beside my house in Manhattan, Kansas, several osage orange trees form their green fruits in late summer and early fall. Their fruits are those little wrinkled globes, textured like brains. Children kick the osage oranges down the street in tedium or insolence. Each time the fruits appear—seemingly overnight, there they hang—I remember: The parakeets would eat those, too.

For me, remembering the diet of an extinct bird is not only a recitation of facts, not only some information to present at one point in a book. It is a spectral way to know the land and its history. I may read, therefore, the absence of parakeets by the presence of life literally just beyond my doors. In a way, the trees on which the forgotten parakeet fed are living memorials to its life.

If hope lives not only as a feeling but as a dedication to vision, then hope flames soon enough, some brush of petal color, fruit color, feather color against the dull milky haze of a sky in which we've punched holes. These days hope asks much from us.

Until recently, I never thought much of osage orange trees, those squat medusas of thorny branches with neither the majesty of a cottonwood nor the quaintness of a redbud. But, once, groves of osage oranges harbored Carolina Parakeets. I hold those trees dearer now.

The Ivory-billed Woodpecker

Overleaf: Ivory-billed Woodpecker, by Alexander Wilson.

In Search of the Lord God Bird

CAMPEPHILUS PRINCIPALIS, THE PRINCIPAL LOVER OF GRUBS: THE IVORY-BILLED Woodpecker, splendid recluse of the swamp.

Two of its nicknames announce the awe that this bird inspired—*the Lord God Bird* and *King of the Woodpeckers*. Observers impressed with the huge Ivory-bill would blurt, "Lord God!" For the Ivory-bill measured nearly two feet long, beak to tail, with an imposing wingspan of two and one half feet and a bill about three inches long and one inch wide at the base. This made the Ivory-bill the largest woodpecker in North America and the second-largest in the world, exceeded in size only by Mexico's Imperial Woodpecker. One ornithologist simply called the Ivory-bill "a whacking big bird."

Brightly yellow-eyed and jittery, the Ivory-bill appeared vividly strange, nearly Mesozoic, as it pounded and drilled the rotting trees of shadow-wet forests and jerked its long, white-billed head this way and that, as if it were the alert guardian of the sloughs. The Ivory-bill conveyed a manic glory as it hopped up and down the sides of cypresses and hackberries or launched off into flight. Upon seeing an Ivory-bill for the first time, one naturalist wrote that he felt "tremendously impressed by the majestic and wild personality of this bird, its vigor, its almost frantic aliveness." This, the King of the Woodpeckers.

The plumage seemed elemental: black bodies with white stripes stretching from the side of the head along the neck and down the back; when the bird perched, a patch of white on the wings' trailing edge showed like a knight's bright culet. The female's recurved crest displayed only jet-black, while the male sported a crest of blood red that gleamed in sunlight shafting through the forest canopy.

The realm of the Ivory-bill consisted primarily of the swamps and river bottomland woods within the vast forest that stretched across the deep South, west to eastern Texas and north to Missouri, southern Illinois, southern Indiana and southern Ohio. (The Mississippi Valley contained 25 million acres of presettlement bottomland hardwood forest, of which five million fragmented acres remain today.) Jacob Studer, writing in his 1881 book *Studer's Popular Ornithology: The Birds of North America*, described these woods as "gloomy swamps and morasses overshadowed by dark, gigantic cypresses, stretching their bare and blasted branches, as it were, midway to the skies."

Into such places—at once beautiful and dreary—ventured the melancholic Alexander Wilson, as he searched for birds to describe and paint for his multivolume *American Ornithology*, published in Philadelphia between 1808 and 1814. The transplanted Scotsman, a minor poet and teacher who took up ornithology at the behest of several Philadelphia naturalists, discovered more than three dozen bird species during his wilderness career. (He also played sad songs on his flute.)

Wilson's tale of his Ivory-bill is among the most amazing stories ever told by a pioneering naturalist, and, hearing it in his words, we detect the registers of attitudes toward a species that would live for another century and a half, against all odds, refusing to give up many secrets until it was very nearly extinct. Here is Wilson's story.

The first place I observed this bird at, when on my way to the South, was about twelve miles north of Wilmington in North Carolina. There I found the bird from which my drawing was taken. This bird was only wounded slightly in the wing, and, on being caught, uttered a loudly reiterated and most piteous note, exactly resembling the violent crying of a young child; which terrified my horse so, as nearly to have cost me my life. It was distressing to hear it. I carried it with me in the chair, under cover, to Wilmington. In passing through the streets, its affecting cries surprised every one within hearing, particularly the females, who hurried to the doors and windows with looks of alarm and anxiety. I drove on, and, on arriving at the piazza of the hotel, where I intended to put up, the landlord came forward, and a number of other persons who happened to be there, all equally alarmed at what they heard; this was greatly increased by my asking, whether he could furnish me with accommodations for myself and my baby. The man looked blank and foolish, while the others stared with still greater astonishment. After diverting myself for a minute or two at their expense, I drew my Woodpecker from under the cover, and a general laugh took place. I took him up stairs, and locked him up in my room, while I went to see my horse taken care of. In less than an hour, I returned, and, on opening the door, he set up the same distressing shout, which now appeared to proceed from grief that he had been discovered in his attempts at escape. He had mounted along the side of the window, nearly as high as the ceiling, a little below which he had begun to break through. The bed was covered with large pieces of plaster; the lath was exposed for at least fifteen inches square, and a hole, large enough to admit the fist, opened to the weather-boards; so that, in less than another hour, he would certainly have succeeded in making his way through. I now tied a string round his leg, and, fastening it to the table, again left him. I wished to preserve his life, and had gone off in search of suitable food for him. As I reascended the stairs, I heard him again hard at work, and on entering, had the mortification to perceive that he had almost entirely ruined the mahogany table to which he was fastened, and on which he had wreaked his whole vengeance. While engaged in taking the drawing, he cut me severely in several places, and, on the whole, displayed such a noble and unconquerable spirit, that I was frequently tempted to restore him to his native woods. He lived with me nearly three days, but refused all sustenance, and I witnessed his death with regret.

Perhaps also regrettable is that Wilson—a keen observer of birds—can hardly be called much more than a decent artist. His work, claimed one of his chroniclers, "was frequently childish and wretched." Unable to draw anything but the most static profiles, Wilson might have accomplished far more had he confined himself to his very real strengths of field observation and description. It seems that Wilson's imprisoned Ivory-bill suffered and died for what is, in fact, a merely competent illustration.

Painting the Ivory-bill stayed utmost in Wilson's mind, but he gleaned from the experience a powerful story that pits the feisty dignity of the proud, wounded bird against the restless encroachment of pioneering settlements, embodied in the fact that it was a *hotel room* in which this Lord God Bird died.

Early on, Wilson and the more successful ornithologist and painter John James Audubon recognized and recorded the obvious and rudimentary elements of the life of this species. They and other naturalists knew of the Ivory-bill's preference for deep forests. They had a sense of its southern range across the country. They knew its call—a strange tinny *yent* or *pait* compared again and again to the notes of a toy trumpet or clarinet, and the loud double rap of the bird hitting its bill against a tree. They speculated about its nesting habits. And they noticed: When the forests vanished, so, too, did the Ivory-bill. The appearance of democracy's entrepreneurial pioneers triggered, it seemed, the disappearance of the King of the Woodpeckers.

Yet even in this matter—evidently straightforward—nuances escaped notice. Did the woodpecker persist near settlements where at least *some* forest remained? How much forest did an Ivory-bill need, after all? Did it live only in mature, old-growth forest or could the bird live in second-growth habitats? Apparently nonmigratory, Ivory-bills seemed nonetheless to roam in search of food, but how far would they go? What of their courtship rituals? Did mates pair for life? And what of their enemies?

Common in some forest locales, apparently rare in others, the Ivory-billed Woodpecker decreased in numbers (whatever they were) throughout the nineteenth century. Such a discernible decline only stoked interest in finding specimens, of course, whether for science or profit, and so furthered the species' scarcity. Once collectors found Ivory-bills clustered in a patch of swampy forest, it was easier to shoot than to observe them at length.

Despite pronouncements that the Ivory-bill was going extinct—then despite pronouncements that it *was* extinct—the species held on, unbeknownst to America's amateur and professional ornithologists. Its life largely had been a secret to naturalists and its apparent death a premature conclusion.

None of this became clear until well after the turn of the century, when the Ivory-bill, so everyone thought, existed only in the cabinets of natural history and on the pages of Wilson and Audubon.

In 1924, for a moment, everything changed. In spring of that year, Cornell University's Arthur Augustus Allen—a distinguished ornithologist and popularizer of bird-watching—was traveling with his wife, Elsa, in Florida. A guide named Morgan Tindle transformed their lives and the course of American conservation when he pointed out one particular nest to the Allens. *A nest of Ivory-billed Woodpeckers.* The astonishment of this rediscovery—almost like a resurrection—soon turned to bitter disappointment after the Allens briefly left the nest. Two taxidermists, apparently with permission from the State of Florida, shot the nesting Ivory-bills. In his articles on the Ivory-bill in *The Auk* and in *National Geographic*, Allen does not discuss his personal reactions to this tragedy. Though slow to anger and almost always optimistic, Allen still must have felt the loss sharply. Even with Elsa to comfort him, he wondered if they had seen the last two Ivory-bills in the world, now irrevocably gone.

The Ivory-bill suddenly had returned, then departed almost as quickly. The sighting, however, raised the possibility that other Ivory-bills might yet be found in some remote part of the deep South, in some swamp that had escaped the sharp blades of progress. Would there be—could there be—another resurrection of the Lord God Bird?

In April 1932, a Georgia Forest Service employee named Charles Newton Elliott believed he could answer that question. He declared the reports he'd received of Ivory-bill sightings in the Okefenokee Swamp to be credible. Elliott's assertion in the magazine *American Forests* stunned some readers. After all, eight years had passed since Arthur Allen's surprise in Florida, during which time no more Ivory-bills had been found or, at least, verified. Besieged by dismayed and unbelieving attacks, Elliott went into the field himself, searching the then-drought-stricken Okefenokee. Though he must have investigated with that peculiar combination of dejection and determination that seems to mark all such searches for a vanishing creature, he found nothing worth reporting.

That same month, apparently on April 15, a Louisiana state legislator and attorney named Mason D. Spencer, a man with a penchant for bars and gambling, raised his gun, looked through the sight and squeezed the trigger. He collected his specimen—triumphantly (and legally, for he had a permit)—then prepared it for safekeeping.

Officials in the New Orleans headquarters of the state's conservation department had scorned verbal reports of Ivory-bills along the Tensas River in Madison Parish, joking about the quality of moonshine available there. These officials could hardly believe their eyes when they found themselves looking at Spencer's specimen: a freshly killed Ivory-billed Woodpecker. One version of the tale claims that Spencer drove south across the state in order to personally deposit the woodpecker on the conservation director's desk. Less dramatically, he may have just mailed the bird.

In either case, the director of Louisiana's Fur and Wildlife Division (the wonderfully named Armand P. Daspit) promptly banned the issuance of any permits for collecting Ivory-bills where Spencer had found his bird. Daspit also directed a game warden who lived by the Tensas River, J. J. Kuhn, to protect whatever population remained in Madison Parish. What had happened in Florida would not happen in Louisiana.

A story asserts that the father of one George Lowery took George to

Spencer's camp, where they heard, then saw, the woodpeckers. Lowery, who would become a well-known ornithologist, gave a talk about this experience at the 1934 American Ornithologists' Union meeting, attended by Arthur Allen.

The finding lured Audubon Society president T. Gilbert Pearson and Louisiana ornithologist Ernest G. Holt—then the director of Audubon's refuges program—to Madison Parish. Between May 12 and 17, 1932, on land owned by the Singer Sewing Machine Company, they saw Ivory-bills. It was in the Singer Tract—a place Pearson called "a great forest"—that the King of the Woodpeckers had persisted in all its wild splendor. There could be no doubt now: The Ivory-bill was back. If the species was no longer a ghost, questions haunted those who yearned to see the bird, to learn of it. Could the Lord God Bird remain with them long enough to reveal mysteries of its life and, therefore, long enough for scientists to understand how it might be protected?

The momentous rediscovery helped launch a 15,000-mile scientific expedition, the likes of which had never been seen before. At Cornell University, the hardworking Arthur Allen considered the news of the Ivory-bills in Louisiana's Singer Tract. By 1934, he also began to wonder about his forthcoming sabbatical. What should he do with a spring semester off from teaching? The decision was more than personal because Cornell, in those days, was it—the one university with a Ph.D. in ornithology as a specific discipline. What "Doc" Allen decided to do would affect not only his career but the university's program as a whole. He could focus on purely scientific research or else conceive of a project that simultaneously would advance the cause of science, help his students and widen public knowledge.

Only a few years before, a film studio had sought the help of Cornell faculty in recording birds; the studio wished only to demonstrate the excellence of its equipment for commercial purposes by capturing hard-to-record bird songs. For Allen and others at Cornell, however, the studio's idea prompted an abiding interest in an innovative form of fieldwork: filming the moving images and recording the sounds of birds in their natural settings.

Refining the studio's techniques and technology, and dipping into his own pocket for the funds to buy expensive equipment, one of Allen's research associates, a stockbroker named Albert Brand, had just published a highly successful book called *Songs of Wild Birds*, which came complete with a phonograph disk of those songs. (Eventually Cornell's Allen would release similar popular recordings.) One night, the investor-ornithologist and the professor talked over Allen's sabbatical options for the coming year. The two men may have met at the beloved home of the Cornell Laboratory of Ornithology, McGraw Hall, an odoriferous, fire-prone and disheveled rabbit warren of a building. Allen and Brand may have conversed about the professor's spring 1935 sabbatical above the sounds of frogs croaking in metal buckets and the voices of graduate students discussing research problems or impending double dates. So thin were the walls that virtually anything said could be overheard. Once, women hired to clean museum specimens were for days overheard discussing a variety of intimate sexual matters; the entranced graduate students could not bring themselves to ask the ladies to quiet down. It's possible that Allen and Brand were heard by a promising student named James Tanner, who, in years to come, would write the definitive natural history of the Ivory-billed Woodpecker.

Old McGraw was the kind of place where graduate students temporarily between lodgings slept on lab tables and hosted dinner parties among cabinets of chemicals. In Old McGraw, escaped snakes terrorized unsuspecting secretaries. It was the kind of place where great ideas took hold, where, perhaps, Brand proposed that Allen record the sounds and film the behaviors of America's rare birds, such as the Ivory-bill. And the goal widened: The ornithologist should record as many birds as possible in a year's time across the entire country. Allen found the idea compelling. Not only could common birds be recorded—some for the first time, including species of the American West—but the threatened birds might have their moving images and their voices preserved for generations to follow. No species then living in America should go unrecorded.

In the 1930s this daring approach to ornithology was essentially untested. True, Brand had captured the sounds of relatively accessible birds for his book—including the Wood Thrush, the Northern Cardinal, even the Bald Eagle—and he had given talks on the subject. But now Allen began to plan an unprecedented venture that would plunge him and his assistants into wilderness they could barely imagine still existed in the United States. The bold project would take bird-recording from the relative comfort of nearby woods and fields and send the ornithologists into places they had never been.

The professor presented the goal of the project in words that marked a clear break with the old, shotgun school of ornithology, which, while on the decline, had not entirely disappeared. His concise summary stated, in fact, a kind of challenge. The Cornell team

> would leave guns at home and would "shoot" the birds with cameras, microphones, and binoculars; [the] object: specimens of bird voices preserved on film, with such photographs, motion pictures, and field observations as would elucidate the habits and appearance of the living birds and determine better methods for their conservation.

In this passage, Allen deemed shooting an inferior form of "collecting"; the whole point of the fieldwork lay ultimately in preserving the birds in their habitats. Such a conservation ethic—still developing among scientists and the public alike in the first half of the twentieth century—offered at least the possibility that we could extend the lives of threatened species. Ornithology no longer would be only descriptive; it would be prescriptive, pointing out the needs of birds and how we could help meet them through enlightened wildlife management.

This fieldwork also would help illuminate the whys and wherefores of bird song. Singing allows males of one species to sort out territories needed for successful courting, mating and breeding. For example, a male House Wren sings his way to dominance over a desired locale, the song communicating to other male House Wrens, in essence, "Keep out." Ornithologists knew that songs helped individuals attract mates and "stay in touch" with members of a migrating flock. They knew that various calls—sounds not as fully phrased as a territorial or courtship song—communicated behavioral responses such as fright. But regional variations in the songs of a species, the linkages between singing and body postures, song durations in different weather conditions—these and other matters needed study. Allen's expedition would put advanced technologies in the service of answering such questions.

The team received technical assistance from engineering professors and financial support from Cornell, the American Museum of Natural History and Brand himself. Then, in February 1935, Allen's team pulled out of Ithaca, New York, setting out across a country gripped by depression and increasingly concerned about a short, bellicose chancellor in Germany.

The affable and unpretentious Allen led the Brand–Cornell University–American Museum of Natural History Ornithological Expedition with ornithologist Dr. Peter Paul Kellogg serving as assistant. The dashing Kellogg would be responsible for the expensive sound and film equipment and would oversee recording. (Kellogg had been involved with the first collaboration with the film studio in 1929 and later would help develop the first portable tape recorder.) Graduate student James Tanner would perform many of the more menial tasks, including assembly and disassembly of the equipment. Brand was not with the team when it left New York.

The men utilized ingenious gear, especially the "sound mirror," a primitive version of the parabolic dish microphone (a technology still used today in capturing, for example, the sounds of live sporting events). Brand had used the sound mirror in his own recordings; this device centered desired bird song onto a microphone, magnifying the sound and excluding unwanted noises (among them, said Brand, was "canine profanity"—barking). The parabolic sound mirror—complete with mic and a small aiming device—made it possible to record from a greater distance, thus reducing disturbance to the birds. Further, the parabolic dish focused difficult-to-record high-frequency songs so that wind interfered less often with recording. The expedition's bulky sound-and-film gear included microphones, headphones, amplifiers, cables, cameras, replacement parts and batteries.

The scientists employed this equipment for what was then called the "sound-on-film" or "movietone" system of sound recording. Lacking magnetic audio tape and digital technology, of course, the ornithologists used sound, light and movie film to capture bird songs and calls. Brand explained the process in his book: "The 'movietone' method of recording consists of transforming the vibrations that affect the microphone first into electrical energy and then into light. The varying intensity of this light is then photographed on sensitized film." From the bird's syrinx to the microphone, then to the amplifier, the sound traveled in waves. Then the signals transferred to "a glowlamp," which fluttered in intensity in response to the sounds. ("[Y]ou are watching sound," Brand wrote.) Forty-five feet of film was exposed every 30 seconds in a sealed chamber with the glowlamp, so that the light flickers—the visual representation of sounds—reacted with the film.

Later, with the film developed, the light pulses could be converted back to sound using a photoelectric cell, an amazing technology for the time. One could simply listen to the sound or channel it to a record cutter that, with a needle, cut peaks and valleys into wax—the first step in the process of making a record or "phonograph" disk. A somewhat cumbersome system, but it worked.

The Cornell team drove two vehicles: a small truck that contained recording and playback equipment and a larger truck that included a tall observation platform, which could be disassembled into a carrying case atop the truck roof. In just 10 minutes, the girded tower could be raised, putting a team member nearly two dozen feet above the ground. This truck also carried the cameras and the camping gear.

The scientists first drove south to Florida, where 11 years earlier the Allens had seen a pair of Ivory-bills. From March 1 to March 23, Allen, Kellogg and Tanner stayed with Albert Brand, who was in ill health, at Winter Park, Florida, according to Nancy Tanner, James Tanner's wife. Keen to various bird sounds, the Cornell crew recorded the calls of the familiar Northern Cardinal and Northern Mockingbird, beginning at or before dawn, so that "milk trucks, tractors, roosters, hounds, and innumerable other sounds of civilization" would not spoil the recordings. In Florida, they filmed exotic quarry, such as limpkins and Crested Caracaras. But no Ivory-bills.

In Tallahassee, Doc Allen, Kellogg and Tanner met up with the newly named curator for Cornell's bird-specimen collection, Dr. George Miksch Sutton, who had for years been known as an excellent bird artist. He also had talents of a more social kind. Sutton, a friend remembered, was "an entertaining conversationalist and story-teller . . . able to sing and to play the piano with gusto." Sutton would leave the others once the men found an Ivory-bill. If they found an Ivory-bill. The expedition would need all of Sutton's songs, all of Allen's puns and limericks and his quiet leadership, all of Kellogg's recording know-how and all of Tanner's youthful enthusiasm. It would be a very long and very challenging journey.

In March 1935, the team stopped in Georgia at the Stoddard Plantation to record Wild Turkeys, then headed back to northern Florida. In early April, the expedition trucks pulled into the only town in Madison Parish, Louisiana, that had electricity—tiny Tallulah, due west of Vicksburg. There the men conferred with Mason D. Spencer, the attorney who had shot the Ivory-bill in Madison Parish three years before, and with the game warden, J. J. Kuhn, who later would aid Tanner in his Ivory-bill fieldwork and who is remembered today, years after his death, as a "remarkable" and "marvelous" woodsman.

Allen, Kellogg, Tanner and Sutton crowded into Spencer's law office in Tallulah and studied the maps of the area spread before them. Spencer spoke of wolves—more numerous there, he claimed, than anywhere else in the United States—and of panthers and black bears. A Southerner born and bred, Spencer cautioned the visiting Yankees about mosquitoes and the ease with which one can get disoriented in the forest bayous of the Tensas River. And, no doubt, he corrected their pronunciation. It's the *Ten-saw*, not *Ten-sas*. "The talk," recalled Sutton, ". . . kept us on the edge of our chairs. There could be no doubt that we were in a fearful and wonderful country."

Spencer also spoke of "Kints"—the local name for Ivory-billed Woodpeckers—and bristled when Sutton pressed him on the matter. Sutton worried, as did the others, that Spencer could be misidentifying the common and widespread Pileated Woodpecker for the somewhat similar Ivory-bill. Although the Pileated presents only black on its back and not the characteristic white patch of the Ivory-bill, one could mistake the Pileated for the Ivory-bill in uncertain light and especially if the Pileated was a partial albino.

But not Mason D. Spencer.

"Man alive! These birds I'm tellin' you all about is Kints!" Sutton recalled Spencer saying. "Why, the Pileated Woodpecker's just a little bird about as big as that." Spencer used his fingers to show a tiny bird, though the Pileated was in fact only somewhat smaller than the Ivory-bill. Spreading his arms, Spencer yelled, "And a Kint's as big as that! Why, man, I've known Kints all my life. My pappy showed 'em to me when I was just a kid. I see 'em every fall when I go deer huntin' down aroun' my place on the Tinsaw. They're *big* birds, I tell you, big and black and white;

and they fly through the woods like Pintail Ducks!" The black-and-white plumage of the wings and tail of the Ivory-bill made the bird look narrow in flight, just like the sleek Northern Pintail duck.

This convinced the Northerners that Spencer knew exactly what he was talking about. Incidentally, Mason Spencer predicted in April 1935 that Governor Huey Long would die violently in the Louisiana state capitol; five months later Long died from an assassin's bullet.

The crew drove into the Singer Tract swamp, but the Cornell trucks, one of them weighing 1,500 pounds, weren't going too far. The unpaved roads were, wrote Allen, "gumbo"—infamous tire-sucking mud created by spring rains. March 1935 had been wet and warm in Tallulah, with nearly nine inches of rainfall. More than three inches had fallen on March 6 alone.

So a hired hand named Ike loaded clothes and some camping equipment from the trucks into a farm wagon. Kellogg may have stayed behind with the trucks or searched by canoe for Ivory-bills, but the others rode in the mule-drawn wagon to J. J. Kuhn's big cabin right between Methiglum Bayou and John's Bayou on a trail so wet it was, Sutton suggested, "practically a lake."

The soil in the Singer Tract is clay and, because the water cannot easily penetrate the tough earth, the area is prone to flooding. Even a few inches of elevation change can affect pooling of water and therefore determine what types of trees and plants will grow. On slightly higher ground, settlers had cleared the old-growth sweet gums and Nuttall's oaks for agriculture. In the early 1860s, Madison Parish was producing more than 100,000 cotton bales annually. The Civil War changed that. Plantations were destroyed and abandoned, and cotton production fell by more than two-thirds. Second-growth forest rose up on those higher lands and former cotton fields, creating a mixed-age forest where old growth remained mostly in the lowlands. Because of climate, though, the second-growth trees grew quickly. In the Singer Tract, one visitor wrote, "It is often difficult for the untrained eye to distinguish a virgin from a second growth forest . . ."

The trail led into a land, a wilderness, that captivated the scientists. "White spider lilies were blooming everywhere. Hooded and Prothonotary Warblers were singing," Sutton wrote. "Since it was early April, and not yet 'fly season,' the mosquitoes did not bother us much." The men arrived at Kuhn's remote home and put down bedrolls on the screened porch, while Ike's son, Albert, set to chopping wood. Surrounded by mossy sweet gums, maples and cypresses, the men enjoyed Kuhn's personable company, excited by his promise that they would soon see Ivory-bills.

The next day Kuhn took Allen, Sutton and Tanner down a wet, dark road, then onto drier land over to John's Bayou. The men marveled at wood chips and scaled bark almost a *foot* long—a certain sign of the Ivory-billed Woodpecker. With temperatures in the 80s, the men spent 36 hours on foot (and perhaps in canoes) as they looked for the Lord God Bird. They waded water as high as ankles, as high as shins, as high as knees and hips. They stepped carefully on wet logs. They shimmied through cat's claw and poison ivy. They walked among old-growth trees whose canopies shaded the understory, creating almost park-like settings clear of brush and briar. Mindful of the wolves, panthers and bears roaming throughout the thousands of acres of forest, the Cornell men neither saw nor heard Ivory-bills. Kuhn worried out loud that they might think he was responsible for, in his words, "a put-up job."

On April 6, 1935, Tanner, Sutton, Kuhn and Allen walked the Singer Tract again, arrayed within shouting distance. (Kellogg was in Tallulah that day, according to Nancy Tanner.) With the spring air warming to its high of 77 degrees, Kuhn suddenly heard a Kint—and yelled and yelled to the others to listen. To the woodsman's dismay no one else had heard the bird. Could they believe Kuhn? "His face was a study," Sutton wrote, "for he was transported with boyish joy." Sutton decided to stick with Kuhn and hope for the best.

Walking along a huge downed cypress, the two men stepped gingerly on its slick surface. "There it goes, doc! Did you see it?" Kuhn yelled,

Male Ivory-bill at nest in John's Bayou, April 1935. Photo by James Tanner.

whirling Sutton hard, trying to orient him to the bird. They almost fell as Kuhn pointed. "It flew from its nest, too, doc! What do you think of that! A nest! See it! There it is right up there!" Sutton *had* barely seen a dark shape fly, but now he had a clear view of a large nest hole. *An Ivory-bill nest* high up a tall tree. Sutton could hardly believe the find. So giddy were the two men—one a sophisticated Ph.D., the other an uneducated woodsman—that they danced and laughed, holding on to each other's arms and shirts to keep from tumbling to the briars and muck below.

Racing through the woods, Tanner and Allen arrived, so taken by joy that they, too, laughed victoriously. They shushed each other, for soon

Shift change at the Singer Tract nest, April 1935.

they heard the call—"strange, bleatlike," Sutton said. This was their prize: the trumpet call of the King of the Woodpeckers, a sound entirely new to them, a sound never-before recorded. Then, as if a novelist were playing tricks, they heard another sound, loud and frightening: A waterlogged tree snapped and snapped and creaked, then slammed in a crackly tumult to the ground.

The following day, Sutton realized his goal of drawing the Ivory-bill directly from life in its own home. Allen and Tanner took still photographs. With no biting insects to speak of and no rain falling, the men lived comfortably in a day of moments, a wilderness miracle. Later at dinner, and after, sitting on Kuhn's porch, they discussed plans. Sutton would depart, while the rest would load cameras and tripods, sound mirrors and microphones, into Ike's wagon. They'd move into the wilderness and live with the Ivory-bills as long as they could and as near as the birds would let them. The nest in John's Bayou was comparatively close—only seven miles from the nearest decent road, "which, in an unbroken forest 18 miles wide and 30 miles long, was indeed fortunate," commented Allen.

In Tallulah, they met with the sheriff and the mayor, and gained permission to pull the trucks into the large jail yard on Green Street, directly across from the courthouse. The men may have slept in the trucks, but, after three days of roughing it, probably chose more comfortable accommodations. Next door, right on the corner of Green and Cedar, stood the clean, moderately priced Post Inn Hotel. A Madison Parish writer and historian, Geneva R. Williams, believes that's where the Cornell team stayed.

Soon enough, curious prisoners watched as the Yankee strangers worked in the jail yard, removing cameras, microphones, film canisters, bedrolls and food from the trucks, then loading all of it carefully into a weather-beaten farm wagon. The loaned wagon, the scientists believed, could handle the muddy trail and deliver them to the nest. This operation, Allen recalled, "furnished much amusement to the inmates . . . several of [whom] volunteered confidentially . . . that they could show us more of these 'pecker-woods' if we could arrange a leave of absence for them . . ."

The next day, after discovering that four mules—not two—had to pull the load, the men set out through shin-high water, Ike's wagon creaking under the weight while Allen and Albert rode on the two lead mules and Kellogg and Tanner walked behind.

They stopped about 100 yards from the nest, lofted 43 feet up the trunk of a dead maple. Ivory-bills excavated nests high in a variety of dead or dying trees, but seemed to use only those that stood in water. Such was the case here. It was April 9, 1935, a cloudy day. The temperature had cooled into the 60s and would stay so till the expedition left. Ike, the mule driver, and Albert took the four beasts away, leaving the wagon with all the equipment. The ornithologists felt a bracing isolation. This was wilderness. Allen called it a "jungle." The forest vibrated and echoed with the songs of warblers and wrens.

Playing with the Ivory-bill's scientific name, *Campephilus principalis,* some punster, probably Allen himself, dubbed their new home in John's Bayou . . . Camp Ephilus. Beneath spring's new foliage, the men cut thick palmettos to cover wet ground, placing the leaves atop the massive roots of an old-growth oak, Allen wrote, in order "to keep our blankets out of the water," though little rain would fall during their stay. They used sticks and branches, in addition to ropes and metal poles, to brace up their big canvas tent. They set benches and mosquito netting amid the previous year's downed leaves of oak and yellow poplar. At this swampy field station, in their long pants and long-sleeved work shirts, they unfolded a tripod with all-important 24-power binoculars, right beside a low-slung chair. Soon one of the team sat down and began the first shift of daily surveillance of the entrance to the cavity's nest. They trained the sound mir-

J. J. Kuhn and Paul Kellogg with "movietone" equipment in the old wagon.

ror with its electric ear at the hole and even raised a blind on a board high into an adjacent tree 20 feet from the dead maple in order to make observations at the same level as the nest.

Allen, Tanner and Kellogg observed for five days, through April 14, joined at times by J. J. Kuhn. Tanner also spent time at Kuhn's cabin, scouting the area for more Kints. Up to five people lived in camp—Allen, Tanner, Kellogg, Kuhn and, perhaps, Mason Spencer or another local sportsman. At least one of them, Kuhn, was armed, a precaution against bears, 'gators and collectors. (When T. Gilbert Pearson had publicized the 1932 rediscovery of Ivory-bills, he made the mistake of telling newspapers that the skins were worth $1,000 each, enticing collectors to shoot some of the rare birds.)

Not only did the scientists tend to the usual camp chores of cooking and clean-up, they had to take precise field notes of behavior at the nest, photograph the Ivory-bills, record their calls and bill-drumming on trees, and capture, if possible, moving pictures of the birds. The men may have quietly hummed a song Sutton had taught them on Kuhn's screened porch and likely stifled laughter at some ill-timed joke delivered deadpan by Arthur Allen.

Allen's prose brims with the excitement of the experience. He must have thought often of the disappointment in Florida, where the two Ivory-bills were murdered just after being found. Here, in Louisiana's Singer Tract, the birds were safe and *right there*:

> The brilliant scarlet crest of the male, the gleaming yellow eye, the enormous ivory-white bill, the glossy black plumage with the snowy-white lines from the head meeting in the glistening white of the wings, are as vividly pictured in my mind as if I were still sitting on that narrow board in the tree-top, not daring to shift my weight, but feeling the board gradually bifurcating me with wedgelike efficiency.

Unlike Alexander Wilson, Arthur Allen could view "his" Ivory-bills in their natural setting.

The men watched each morning when, at 6:30, the male (whose duty it was to spend the night incubating the eggs) "tapped on the inside of the nest hole," then "grew more impatient [and] stuck his head out," calling "a few 'yaps' or 'kents' in no uncertain tone." When the female arrived, "a little intimate conversation then ensued and she entered." The only foreboding note to this otherwise fetching daily ritual happened when the male would spend several minutes preening and scratching, "as if he were infested with mites," Allen wrote.

Until 4:30 each afternoon—when the male entered the nest for his long 14-hour night shift—the parents alternated between incubating and feeding in two-hour intervals. James Tanner described in wonderful detail the "regular ceremony" that usually took place when a shift concluded:

> One would fly or climb to the entrance and signal the other by calling or pounding. We occasionally heard the setting bird answer by pounding on the inside of the cavity. It would flip out of the hole and catch itself on the outside of the tree beside its mate, and the pair would then exchange a low, almost musical call that had a conversational quality, often given with their bills pointed upward.

Despite a reputation for extreme wariness, the birds "quickly, rather surprisingly so, accepted the presence of from two to five people, the blind [in the tree], and the microphone with its large reflector . . . ," wrote Tanner. The birds continued with their ceremonial shift changes.

Presently, though, the team had a decision to make. The observations of the Ivory-bills were going extraordinarily well; the team had even managed to capture film of an Ivory-bill at the nest entrance. Should they stay until the young bird hatched and fledged? Or should they press on? One of the other species they wished to film, the Lesser Prairie-Chicken, had begun its strange mating rituals on the shortgrass plains to the west.

They decided to leave and return later to check on the Ivory-bills. It proved to be a fateful decision. If Allen felt any trepidation, he may have kept it to himself, and, after all, J. J. Kuhn would guard the nest as best he could.

After Ike, the mule driver, arrived at Camp Ephilus at noon on April 14, the men loaded their gear and slogged out. The temperature rose to nearly 80 degrees, and they sweated nervously, hoping the equipment would not be damaged. Eventually, they arrived safely in Tallulah and spent the next 48 hours reinstalling the film-and-recording equipment back into the two vehicles. The men planned the route toward Oklahoma's Lesser Prairie-Chickens.

The timing could not have been worse. Spring 1935 in the plains of western Kansas, Oklahoma, Texas, eastern Colorado and northeastern New Mexico saw little in the way of crops, unless one counted foreclosures as a kind of harvest. Wheat prices continued to collapse. What filled the wide plains sky wasn't sunlight and frontier optimism about limitless abundance and profits; it was, instead, dust. Nineteen thirty-five became known as the second year of the Dust Bowl, when wind lifted over stunned county after stunned county the overgrazed and overplowed soil, countless dry granules of airborne dirt and sand freed by the plow from the staying power of native grasses. "The blackest year," says environmental historian Donald Worster, "was 1935."

On April 11, in Tallulah, an ominous foreshadowing occurred; the day was "hazy with dust," a weather watcher reported, with visibility down to between a half-mile and three-quarters of a mile. A Great Plains dust storm had reached northern Louisiana. Whether the men in the swamp noticed, no one said in published accounts. On April 14—the day Allen, Kellogg and Tanner had rolled up blankets, stowed pots and silverware, closed their notebooks and carefully packed up the scarred gray wagon in the John's Bayou swamp—townsfolk in western Kansas noticed something odd. "Yards were full of birds," Worster relates, "nervously fluttering and chattering—and more were arriving every moment . . ." What was going on?

The day had not yet been nicknamed, but it would be soon: Black Sunday. "A black blizzard" as tall as buildings in eastern cities rose up on the horizon and rolled in awesome billows through Kansas, Oklahoma, Texas, New Mexico, coating everything—lungs, plates, linens and clothes inside shut closets—stranding drivers and passengers, burying shacks and fence lines, even collapsing roofs. In Meade County, Kansas, 33 people died.

Reports of the dust storms forced the team to detour to E. A. McIlhenny's private egret and heron refuge, also in Louisiana. After spending three days there, the Cornell scientists hit the road and still arrived in Oklahoma in the midst of the vast, otherworldly dust. The wind-driven dark scrims battered their windshields and buffeted the trucks. They must have questioned their decision, wondering whether it would have been more fruitful to stay in the Singer Tract. After all, the crucial fledgling phase and interactions between young and parents remained a mystery, and they risked missing such observations.

But the expedition continued west, hoping for a later return to Louisiana. The team reached the Verne Davison Ranch a few miles north of the Canadian River. For 36 months, that part of Ellis County hadn't seen any rain to speak of. The team stayed for eight days, wishing for a break in the relentless howl of wind and dust. On the last day, it happened: The wind calmed enough to allow filming and taping of the Lesser Prairie-Chickens. Those birds weathered the Dust Bowl perhaps better than the Okies who fled land made barren by overconfidence. And the prairie-chickens fared at least as well as the ranchers and farmers who—with government assistance—hung on long enough for the drought to break and for soil conservation practices to be accepted.

As the Cornell team drove next toward Colorado, they remarked on some of the strange things they saw in the plains. The dust, of course, but also the cows on the Davison Ranch that ambled up to the green-painted camouflage of their blind—"greener than anything in that whole country," said Allen—and who tried to graze and nibble the sides of the blind, as if it were forage. When finally rain arrived, the cattle lumbered through newly formed miniature puddles and small, temporary lakes to sip methodically from the windmill-pumped drinking troughs to which they were accustomed. Next, the men drove across, in Allen's words, "the barren, wind-swept prairies of western Kansas into the verdant irrigated stretches of eastern Colorado. . . ."

I like to imagine Allen, Kellogg and Tanner driving through my state, through Kansas, through the worst of the Dust Bowl's haggard territory, with their captured light, the virtual representations of the Lord God Birds. Films of Ivory-billed Woodpeckers, stowed in dark steel canisters, vibrated with the hum of wheels and axles.

Back in the Singer Tract, J. J. Kuhn watched the Ivory-bills with increasing worry. In late April, he had seen disturbing behavior: The adults were not feeding any young. Rather, they diligently, obsessively preened their feathers. He reported this to Arthur Allen upon his return to the Singer Tract on May 9, 1935. Allen found the nest in John's Bayou abandoned and the cavity littered with wood chips and broken bits of eggshell. Having satisfied themselves that the parents were not returning to the former nest, Allen, Kuhn and the National Park Service's L. M. Dickerson cut the tree down, examined the cavity and emptied what detritus remained into a bag.

In Tallulah the next day, perhaps back at the Post Inn, they turned on a light and placed some of the wood chips beneath it. "The material gradually came to life and swarmed with innumerable tiny mites and we soon felt them crawling all over our hands. We quickly bagged the entire contents again, trying to return as many of the creatures from our hands as possible," Allen and Kellogg wrote. The eerie scene clearly suggested the cause of the nesting failure.

Later that day, Allen observed a second nest in Mack Bayou, where the parents were feeding their young. This nesting also failed within a few days; yet, upon cutting down the tree, the scientists found *no* mites in the cavity.

Could a predatory Barred Owl have been responsible for the Mack Bayou nesting failure? Perhaps, though no signs of blood or struggle appeared evident. Upon vacating a nest, young Ivory-bills would roost on boughs or on the sides of trees. For whatever reason, they never used former nests for their nightly roosting throughout the year. A young Ivory-bill was thus exposed for several nights until it had excavated its own roosting hole. Recently fledged birds may have been vulnerable to owl predation.

Certainly there was one predator who could have taken young Ivory-bills without a bloody struggle—humans. Locals may have wanted to capture or kill the young birds in order to saw off the ivory-colored bill, which attracted interest as a valuable charm or curiosity. There can be little doubt that collectors were paying hunters to secure specimens. Later, there were even rumors that E. A. McIlhenny himself—who had hosted the Cornell team—had paid someone to collect Singer Tract Ivory-bills for him in 1934 or 1935. Audubon staffer Richard Pough wrote in 1943, "I suspect more birds than we realize have been shot around here just out of curiosity . . . The head of the Tallulah State Police told me about his never believing there was such a bird, until [someone] brought in a pair to show [a] gang of hunters . . ."

What had happened to those nests where mites could not be blamed for the abandonment never could be determined. Not even the bodies of the young birds were found. Perhaps the parents had removed the nestlings after they died.

The mysteries of these nesting failures troubled the ornithologists, but they had many birds yet to record and film. So the expedition continued. In Utah, strong mountain winds often delayed the team's recording efforts. James Tanner and Paul Kellogg did not complain. They slept in past Allen's 4:30 A.M. alarm. (Allen was leading a predawn field class.) And the team did manage to capture the calls of the American Dipper, which walks underwater; the chunky bird actually perched on the crew's streamside microphone. In Montana, the Cornell team recorded the endangered Trumpeter Swan, which has since been rebounding in number. That was their last task, but not their last adventure. They next drove almost nonstop to Ithaca—and arrived only two hours ahead of a flash flood. Arthur Allen must have chuckled as he heard the rain pounding his house up in wooded Cayuga Heights: "Our trucks finally became marooned in my own back yard." The expedition had enough motion-picture footage to span 10 miles. The team had recorded more than eight dozen bird species. It had been a stunning success.

Back home, Allen told many tales of adventure. One of his favorites concerned the Colorado hailstorm and the plovers. To Mountain Plovers nesting on the ground—in the shortgrass plains east of Denver—a sudden fall of hail threatened mortal danger. Seeing a plover in distress at the bombardment, which could easily shatter its eggs and even kill adults, the Cornell team drove within inches of the eggs and so sheltered them from the storm. The parent plover quickly returned to its nest.

Suddenly the truck, this giant black intruder, became a haven, not only to the Mountain Plovers but also to a small flock of Lark Buntings that quickly flew beneath the chassis. "Then came a Western Meadowlark," wrote Allen, "pitifully frightened."

He longed for the shelter of the truck, but he was a timid bird and each time he approached within ten feet . . . and could see us inside, his courage deserted him and he ran back . . . At last, summoning all his courage, he made a rush for the car and slipped safely beneath with the other birds . . .

The meadowlark, now safe, sang a song of gratitude—loud, clear and buoyant, right below the car.

After the storm passed, the ornithologists looked at the thick clouds above, the black specks of moving birds near and far, and the white concretions of hail below and they wondered at the vast and rolling grasslands that, as with so many places, men had lately changed.

Returnings

Wine-hearted solitude, our mother the wilderness,
Men's failures are often as beautiful as men's triumphs, but your
   returnings
Are even more precious than your first presence.
            —Robinson Jeffers, "Bixby's Landing"

FOLLOWING THE CONCLUSION OF THE BRAND–CORNELL UNIVERSITY–
American Museum of Natural History Ornithological Expedition, James
Tanner returned alone to the forests and bayous along the Tensas River,
back to land named for the company that owned it: the Singer Tract. For
weeks at a time, he lived in a southern jungle, looking for and at Ivory-
billed Woodpeckers. His work formed the basis of his Cornell dissertation
and the first research report of the National Audubon Society. (From
1937 to 1939, that organization spent $4,500 supporting Tanner's eco-
logical studies.) Quite literally, James Tanner wrote *the* book on Ivory-
billed Woodpeckers. In doing so, he found that the rare bird had become
like kin. "I never tired of watching them," he once wrote.

Tanner estimated that only two dozen Ivory-bills still lived in North
America by 1939, just six in the 80,000-acre Singer Tract in Louisiana and
the rest suspected to reside in the Big Cypress, the Apalachicola River bottoms and the Gulf Hammock–Suwanee River area in Florida. Tanner also
wondered if South Carolina might harbor a few of the black-and-white
giants. In retrospect, some latter-day ornithologists have suggested that

Tanner's population estimate was too low; they argue that the number of Ivory-bills then living may have been closer to 12 times what Tanner thought. A Louisiana state forester said in 1939 that he believed nearly 100 Ivory-bills lived in the Singer Tract alone, though that estimate seems unreasonably high.

Nonetheless, by the 1930s and 1940s, the disappearance of the Ivory-bill seemed to be quickening. Clarifying the ultimate and proximate causes for the bird's decline, sorting out the nuances of how the bird depended on wood-boring grubs and forwarding a plan to save the species all constituted central elements of Tanner's labors.

And labors they were, exhaustive and exhausting. Traversing 45,000 miles in his Model A Ford and in trains, as well as "uncounted miles by foot, boat, and horse," the youthful doctoral candidate worked in the winter and early spring to avoid summer's heat and leafy cover. He drove down dusty roads, pulled into isolated towns across the deep South and stopped at sheriff's offices, sawmills and general stores to ask locals what they knew of his quarry. The Cortland, New York, native soon learned to avoid asking directly about Ivory-billed Woodpeckers, which sometimes encouraged replies to entice Tanner into hiring the respondent to serve as a local guide, regardless of the person's expertise. He began to show drawings of the Pileated and the Ivory-bill, calling the bird by colloquial names like Old Kate or Lord God or Woodcock.

Tanner spent eight months of his fieldwork just trying to find Ivory-bills in places other than Louisiana's Singer Tract. He read old accounts of the species. He studied maps. He looked for black-and-white woodpeckers that gave the impression of narrow wings and a slender tail—like the Northern Pintail ducks to which Mason Spencer had once compared the Ivory-bill.

Everywhere he went, Tanner listened for the Ivory-bill's loud, double rap of its bill on a tree and for the resonant, nasal-like *kent*. He and others compared the Ivory-bill's "yap-like" call not only to musical instruments but also to the less powerful call of the tiny Red-breasted

Nuthatch. It's remarkable that two birds so different in size can sound so similar: the Ivory-bill as long as my arm and the nuthatch, a bird I could cup in the palm of my hand.

Tanner even took a saxophone mouthpiece to the Singer Tract to see if its notes would interest the Ivory-bills. Saving a threatened species can involve many tasks, but who would have thought a recital was among them? Amid the sweet gums and cypresses, the White-eyed Vireos and sleeping owls, Tanner "tooted" the mouthpiece—which sounded similar to the bird calls but rather flat. No Ivory-bills arrived to be his audience.

The Santee Swamp and the Pee Dee River bottoms in South Carolina. The Okefenokee Wildlife Refuge in Georgia. Bear Bay, Jane Green Swamp and the Royal Palm Park in Florida. Mobile Delta, Alabama. Yazoo Delta, Mississippi. To add to his confirmed list of six individuals in the Singer Tract, James Tanner looked everywhere Ivory-bills might be living. In some places, he discovered reliable reports but no birds. In others, he found peeled bark—a sign of Ivory-bills searching for food—but no birds. Elsewhere, the forest was too small or already cut. Almost everywhere, people mistook Pileateds for Ivory-bills. But not Tanner. His breath caught each time he glimpsed a large, black-and-white woodpecker. Each time the bird proved to be a Pileated, whose undulant flight differed from the usual straight-and-true manner of the Ivory-bill. He watched the Pileateds fly away, sometimes calling a petulant *wukka-wukka-wukka*.

James Tanner never found a seventh Kint.

So he had to confine his field observations to those birds still remaining in Madison Parish, Louisiana. He would carefully provision himself and pack his gear onto saddlebags, then ride in by horse, or carry his gear on foot. Once settled, he often had little difficulty noting the presence of the birds—all he had to do was listen for their calls—but he sometimes found himself running full speed to catch up with the big woodpeckers just out for their morning feeding. One year, in 1938, he found a nest on February 17, just one day after arriving in the Singer

May 1937. James Tanner beside a large honey locust in the Singer Tract.

Tract to begin his regular winter research. The next year, however, he spent five hard days searching before locating a nest. "My journal," Tanner remembered, "is full of such comments as 'saw old sign [scaled bark], lots of almost impenetrable vines, and no ivory-bills.' " Occasionally, he'd whap a stick on a tree for a response from any nearby Kints. Fortunately, once he had determined locations of the birds' current roosting holes or nests, he could simply wait for the birds to emerge.

Tanner possessed the attributes needed for such work. He had "an ability to rough it and to get along with all kinds of people in all kinds of situations, a natural adaptability, ingenuity, initiative, originality, and a willingness to work," his doctoral supervisor Arthur Allen wrote. "Above all, he had . . . a clear mind and superior intelligence." His wife, Nancy, now 82, tells me from her home in Knoxville that "Jim was a . . . humble man." Friends and acquaintances recall Tanner as reserved yet friendly—and possessing a sense of humor. He was a man who spoke deliberately, as if thinking aloud.

Even with all these laudable characteristics Tanner could not always work alone; he needed help in the deep woods of the Singer Tract. On March 6, 1938, Tanner climbed a tree, grabbed a nestling Ivory-bill out of its hole, banded it and began to trim away a branch that had been blocking the view from the ground. He watched in dismay as the nestling leapt out and "fluttered to . . . a tangle of vines, where he hung, squalling and calling." That day, and on others, Tanner had the assistance of J. J. Kuhn, the Madison Parish woodsman and warden who kept watch on the birds when the scientist returned to New York.

That morning Kuhn—a tanned and powerful man—held the bird, "still squalling loud enough . . . to be heard in Tallulah." Then Kuhn became a perch for the young Ivory-bill, as the frightened bird clambered up to his hat. Tanner took pictures of the nestling Ivory-bill atop his friend's head: "I had snapped six before I realized that I had not stopped to focus the camera," Tanner wrote in one of two unpublished reminiscences that Nancy shared with me. Eventually Tanner got his photos, tucked the nestling into two handkerchiefs, placed the bundle under his shirt and put the bird back in its nest before the parents returned. Kuhn and Tanner nicknamed the bird "Sonny Boy."

Tanner would settle into Kuhn's cabin, which served as a base camp, and begin a regular schedule of predawn breakfasts followed by hikes in search of Ivory-bills. There were days when Tanner spotted panther tracks. One day he watched a beautiful black wolf striding through palmettos. Once, nearly lost, Tanner heard a wolf howling at dusk and, newly motivated, found his way back to camp. After supper, Tanner and Kuhn would discuss the day's happenings and review topography—where they could find dry ridges, foot logs and clearings, where palmettos and vines thickened to create impasses and where ancient trees interspersed with young ones. Owls would punctuate winter nights with their lonely calls, and, sometimes, if Tanner slept outside, he'd see stars caught like white blossoms in the branches of the trees.

Through careful observation, the ornithologist verified that the same behaviors the 1935 expedition had witnessed persisted: Ivory-bills emerged later in the morning than did other birds; they preened, they called, then flew off for feeding forays, rested during the afternoon, then fed as they headed back to their roosts. While discouraged by rain, the birds were otherwise "quick and vigorous, almost nervous," Tanner wrote. "When perched and alert, they have a habit of swinging the body quickly from one side to the other, pivoting on the tail pressed to the tree, pausing to peer back over the shoulder, then swinging back and looking over the other shoulder, at each quick swing flirting the wings." The Ivory-bill seemed to glory in its own body, in brisk, keen movements.

But just as the physicist can alter subatomic particles by the very act of observation, so too did Tanner occasionally modify the behavior of his objects of study. Once, while dismantling a cloth blind high in a tree, Tanner's motions caught the attention of three adult Ivory-bills, "shouting and knocking so nearby, glaring at me from their yellow-irised eyes, while I straddled the crotch of a tree some forty feet from the ground; for the first and only time in my life there were too many ivory-bills too near. I covered myself again with the cloth of the blind." And twice he saw the woodpeckers strike their bills against the metal spikes he'd driven into a tree. Confronted with something they neither recognized nor understood, the birds stayed clear of the strange, bright objects.

The Singer Tract teemed with woodpeckers: Downy, Hairy and Red-bellied. Red-headed. Yellow-bellied Sapsuckers. Northern Flickers. Pileateds. Of course it was not surprising that a deep forest with a wide range of tree species, and trees of varied age, should attract so many members of the *Picidae* family. (Equipped with strong feet, stiff tails and tongues so prodigious they actually wrap around their thick skulls, woodpeckers are marvels of evolution. Many an ancestor's brain must have been addled before the Woodpecker Proper appeared.)

It wasn't the total sum of woodpeckers, but their relative numbers that surprised Tanner. He found that in a six-square-mile area of forest, the "maximum abundance" of Ivory-bills was just one pair. In contrast, *one* square mile of appropriate habitat could support 21 Red-bellied Woodpecker pairs and 6 Pileated pairs. Thus, for every 36 Pileateds you could expect to find 126 Red-bellieds—and 1 Ivory-bill.

The reason hid in the feeding preferences of the species and the amount of habitat available to support those preferences. Ivory-bills spent much time not hammering at trees but *scaling* whole swaths of bark to expose the wood beneath. The birds preferred to scale the relatively thin bark of the sweet gum and Nuttall's oak because those trees, when dead or dying, attracted more fat, wood-boring *Cerambycid* beetle grubs. (In Florida, the birds scaled dead pines and cypresses.) The rest of the scaling was distributed lightly among various species such as honey locusts and elm. Usually the birds whacked away tight bark with gusto, "with sidewise blows and quick flicks of the bill," Tanner said, then the birds voraciously gobbled up the exposed grubs.

The Ivory-bill also chipped and dug into trees to forage and to create nesting and roosting holes. (The bird could dig a five-inch hole in a hackberry tree in fewer than 60 seconds.) Whether scaling or digging, the Ivory-bill showed a predilection for trees between more than 1 foot to 3 feet in diameter—those most prone to decay and infestation by wood-boring grubs. Yet three-fourths of the Singer Tract consisted of trees between 3 and 12 inches in diameter, another factor apparently working against the bird's survival in that limited habitat.

By way of comparison, the plentiful Pileated Woodpecker digs more than it scales and does so in all kinds of trees. "The important difference," Tanner found, "... is that the Ivory-bill [fed] mostly on the borers that live beneath the bark of freshly dead wood, while the Pileated feeds mostly on the borers that live within the sap and heart of longer dead wood . . . "

What all this means is that the Lord God Bird existed in a niche almost as slender as a feather. A specialist in diet, this species clearly favored the larvae of beetles from trees dead for only two or three years. "I do not know why [they] should prefer to feed upon insect borers that are relatively scarce," Tanner wrote, "unless it be that when the borers beneath the bark are abundant, they are very abundant . . . "

The Ivory-bill also ate fruit, such as persimmon and wild grapes, and seeds, such as poison ivy and magnolia, as well as various berries. One analysis of three Ivory-bill stomachs determined that a substantial percentage of the food was vegetable. None of the foods were crops or orchard fruits; the species never became an agricultural pest.

Ivory-bills may have been quite mobile in searching for food sources, but could become attached to a site that yielded ample supplies of wood-boring grubs from newly dead trees. The Singer Tract Kints apparently had become site-dependent. They appeared to forage only within the confines of that great forest, and Tanner observed them to eat *only* their desired grubs found therein. Although infestations of grubs were "eruptive and undependable," as Tanner noted, they were common enough in the Singer Tract to support the remaining Ivory-bills there.

The chain of dependency had become, evidently, too delicate for the species, but only because of the insatiable American appetite for lumber. Logging meant a loss of the kinds of trees this bird *knew* in its blood—feeding trees containing wood-boring grubs, nesting trees, roosting trees. Woodpecker expert Jerome Jackson specifically implicates "rapid" rail development in the South at the turn of the century, which made logging that much easier by opening access to previously impenetrable primeval forests. Rails, blades, saws and pallets, this equipment of havoc and profit apparently doomed the Ivory-bill throughout the South.

James Tanner summarized the situation in his October 1942 Audubon research report when he wrote that:

> the Ivory-bill is present only in forests where dead and dying trees are frequent and other woodpeckers are abundant, conditions which normally prevail only in tracts of uncut, mature timber . . . Mature forests of large, old trees have almost disappeared [in the Southeast], and these conditions favorable for the Ivory-billed Woodpecker will very probably never again prevail.

A verb of contestation—"prevail"—was perhaps more appropriate than even Tanner himself realized as he prepared his study for publication. After all, even if the Ivory-bill had disappeared from, apparently, almost every place in its former range except the Singer Tract, it held on there. "Prevail" relates in its root both "to be strong" and to "value"—and the struggle to study and to save the Ivory-billed Woodpecker in its last-known home was a matter of both, a matter of courage and greed, a reflection of the values of triage conservation and of a particularly recalcitrant kind of capitalism.

———

Preventing the extinction of the Singer Tract Ivory-bills demanded at least a rudimentary understanding of the bird's reproductive behavior and a sense of what other factors had disrupted the lifeways of the species once its habitat began to be destroyed.

Ornithologists believed that Ivory-bills would pair for life and that mates would remain together year-round. Young might stay with their parents for several weeks. After a courtship ritual—the touching of beaks—Ivory-bills would copulate, and the female lay between one and four white eggs sometime between January and May. The male and female shared the incubation duties, which lasted about three weeks.

The loss of a mate made finding a new one that much more difficult, given the limited pool from which to select a replacement. Further, the loss of a mate during the courtship season might mean that a fertile female would not breed and produce eggs—a lost opportunity. The species had a low reproductive rate to begin with.

As Americans cut forests in the nineteenth and twentieth centuries, other things began to interfere with the Ivory-bill's ability to raise and fledge enough young birds to exceed the death rate. The proximate factors apparently most important in the final years of the Ivory-bill's existence included nest infestation by mites, inbreeding and scattered populations, which may have led to a lack of synchrony in sexual rhythms among remaining males and females—a cause of infertile eggs, if any eggs at all. (The European Honeybee used hollow *trees*, not simply holes, and so would not have affected the Ivory-bill, according to biologists.)

Rural poor, living on a subsistence diet, killed Ivory-bills as food and used their bills for trinkets. John James Audubon noted that steamboat travelers at a dock could buy two or three Ivory-bill heads for a quarter; the price over the years would increase to $5 a beak, Tanner wrote, with some folks carving the beaks into "watch charms or fobs." Native Americans also had valued this woodpecker for its striking plumage and its massive yellow–cream-white bill, and they may have reduced some groups of the species in particular areas, though without threatening widespread or permanent depletion.

Collectors shot the woodpeckers as natural history trophies or study objects. Such shooting at the end of a species's life often devastates local populations. A nongame-protection act passed in Florida in 1901 did little to protect the Lord God Bird in that state, though a jury found at least one dealer guilty of peddling dead Ivory-bills.

A further proximate factor in the Ivory-bill's decline may have been competition between that species and Pileated Woodpeckers. Ornithologist Lester Short suggests that as primeval habitat disappeared, Ivory-bills moved "into secondary habitats and became widely scattered . . . " This would have been an obstacle to mating and to raising young, not to mention finding food. How tolerant these two species were of each other, under such conditions, is not clear.

Even with the nesting failures in the Singer Tract, Ivory-bill parents managed to raise 19 young there, between 1931 and 1939. If those birds had had enough room, enough habitat, they might have persisted longer than they did.

So precarious was the bird's existence that an incident involving an Ivory-bill nest worried Tanner and Kuhn for several hours, until they realized what had happened. After Tanner had cut down the stub of a tree containing a former Ivory-bill nest, he heard "a rotund oath coming upwards through the leaves." Jack Kuhn had looked inside the nesting cavity—and found eggs! But the Ivory-bills already had *fledged* their young bird. How could this have happened?

Tanner and Kuhn felt sick. "Do Ivory-bills really lay again, and so soon after young have left, and in the same nest?" they wondered. "What kind of observers were we that we did not know what the woodpeckers were really doing? Had we, through what amounted to criminal negligence, broken up an ivory-bill nesting? We carefully wrapped the eggs in spanish moss and silently returned to camp," wrote Tanner.

Fortunately, the men later identified the eggs as those of a Wood Duck, not an Ivory-bill.

The fact that the Ivory-bills continued to hang on as Tanner visited the Singer Tract in the late 1930s and early 1940s provided some small encouragement. The ornithologist believed the species might survive, but only if the Singer Tract could be saved.

In 1916, the Singer Sewing Machine Company purchased the 80,000-acre forest with its expanse of lush second growth interspersed with fine stands of primeval trees. A few years later, the State of Louisiana began to manage the Singer Tract as a game refuge, but never acquired title to the property. Significantly, the lease did not prohibit commercial exploitation of resources. So, in 1937, the Singer Company sold logging rights to Tendal Mill and to the Chicago Mill and Lumber Company, a firm still spoken of with quiet derision by many in Madison Parish. The next year, in 1938, logging began. (Any land logged by Chicago Mill became that firm's property; the company then leased the cleared land to poor families under terms, some say, that were financially burdensome to the new tenant farmers.)

To save enough of the Singer Tract to sustain a handful of the woodpeckers, Tanner detailed very specific measures that would have preserved habitat for the Ivory-bill and still left lands for the loggers. Having seen a female and a young Ivory-bill adjust to the presence of loggers in 1941, in John's Bayou, Tanner believed humans and Lord God Birds could coexist in the Singer Tract.

In the final pages of his study, Tanner proposed a ban on all hunting, strict supervision of visitors interested in seeing the Ivory-bills, employment of a warden and experiments to determine the best way to kill trees by incisions ("girdling"—meant to attract the Ivory-bills' desired grubs). He also recommended that low-quality timber be left to stand—and be girdled—for the benefit of the woodpeckers. He advocated selective logging rather than clear cutting in order to protect habitat and raise funds both for the loggers and for refuge management.

Tanner outlined three land-management categories: unlogged, primeval "reserve areas" of about 2.5 square miles per Ivory-bill pair, the absolute minimum necessary for foraging, he thought; "partial cutting areas" where "each tree must be judged on its own condition," with dying and dead trees left to stand; and "logging areas" that could be clear cut. In either clear-cut zones or selective-logging stands, timbermen would have to forego the use of tramlines, trucks and tractors—using draft animals instead—and all operations would be under the supervision of the refuge manager.

The Singer Tract could have become a major national park, Tanner felt, but with two-fifths of the wilderness destroyed by the time America declared war on Japan, that goal seemed less obtainable. Protecting what remained in Mack Bayou and John's Bayou, however, might save the last six Ivory-bills, as well as bears, panthers, wolves and other creatures. All else failing, the Ivory-bills could be captured and relocated or kept in captivity. Tanner believed that the way market and plume hunters netted Carolina Parakeets—by covering their roost holes—could serve as a model. Such an operation never took place.

James Tanner saw Ivory-billed Woodpeckers for the last time in December 1941. (The disappearance of the Ivory-bill ironically coincided with the rise of Woody Woodpecker to national fame. Woody's red crest and golden beak look not unlike an Ivory-bill's, though cartoonist Walter Lantz had been inspired to draw his colorful creation after being pestered by a western Acorn Woodpecker.) Three months after visiting the area with his newlywed, Nancy, Tanner wrote, "The Singer Tract is now in the process of being destroyed." In his note of March 1942 in the professional ornithological journal *The Wilson Bulletin,* he provided this grim synopsis:

> There is little doubt but that complete logging of the tract will cause the end of the Ivory-bills there, and since the surrounding country is young second-growth forest and cultivated lands, it will doom the woodpeckers to a vain search for suitable food and habitat. Discussions are being carried on with officials of the company controlling the tract to determine what might be done for the Ivory-billed Woodpecker . . . The future . . . is far from bright, but there is still a chance for its survival if we can plan well enough ahead.

The prose barely holds back heartbreak. Words like "doom" and "vain" are strong words for a scientist to use. Though the professionalizing of diction and the use of the third person largely had replaced the romantic "I" of an Audubon or a Wilson, Tanner could not hide his distress.

From seven pairs and four young in 1934, the dwindling forest in

1939 supported but one pair, one young bird and three males. All but two of them lived in John's Bayou. One male lived in Greenlea Bend. Another frequented Mack Bayou and, just to the northeast, an area known as Titepaper. Hunter's Bend had been logged in 1938, about two miles west of Mack Bayou. Ivory-bills already had abandoned Little Bear Lake and their northernmost habitat in the Singer Tract, the aptly named Bayou Despair.

The Singer Tract would not fall, however, without a fight, one led by the Audubon Society's president, John Baker, who had twice visited the area. Baker appealed to President Franklin Roosevelt, who directed the Secretary of Interior to take the matter under consideration. Soon, the directors of the Forest Service, the National Park Service, the Fish and Wildlife Service and the War Production Board got involved. According to Don Eckelberry—the Audubon Society painter who was the last naturalist to see the last Singer Tract Ivory-bill alive—the chief of the War Production Board believed that the war effort could spare some trees on behalf of the Ivory-bills. Baker wanted Chicago Mill to "waive its contract rights to cut the remaining virgin timber and agree to sell at a reasonable price . . . sections of land in a cutover buffer area."

Baker further secured a $200,000 pledge from Governor Sam Jones of Louisiana for the proposed land purchase. Four governors—Jones, Prentice Cooper of Tennessee, Paul B. Johnson of Mississippi and Homer Adkins of Arkansas—wrote to Chicago Mill asking that the Ivory-bill be spared.

On December 8, 1943, the brokers of a potential land deal gathered in Chicago: Louisiana's conservation commissioner; the refuge director of the U.S. Fish and Wildlife Service and the agency attorney; John Baker; and the chairman of the board of the Chicago Mill and Lumber Company.

The company argued that less logging would mean financial loss and fewer jobs. The firm had suffered setbacks and layoffs during the Depression. But the offer of $200,000 from Louisiana would seem to have mitigated against the company's argument. Furthermore, *German prisoners of war now did the actual logging.*

The company refused to deal. "We are just money grubbers," said James F. Griswold, Chicago Mill's chairman of the board, during the meeting. "We are not concerned, as are you folks, with ethical considerations." Chicago Mill's president, F. W. Schatz, endorsed the hard-line position. Singer Manufacturing Company Treasurer and Vice President John Morton told Baker that Singer didn't care where its money came from, so the decision to log or to sell belonged solely to Chicago Mill. And Chicago Mill would not assist—with any way—with the development of a refuge "unless forced to do so," according to Baker.

Efforts to protect the Singer Tract through Congressional legislation in 1940 had not come to fruition. Baker failed in his attempts to persuade Louisiana Governor Sam Jones, then his successor, Jimmie Davis, to condemn the land. Because the Singer Tract also was a haven for migrating ducks, Baker even sought the intervention of the Migratory Bird Commission, but that board, beset by internal political squabbles, proved to be of no help.

In a last-ditch effort to assess the situation on the ground and perhaps to find other, nearby forests that might be available for purchase, Baker sent Audubon staffer Richard Pough to the Singer Tract. (Pough would go on to a distinguished career in conservation, including, among many achievements, the presidency of The Nature Conservancy.) Baker especially hoped that Pough would find an Ivory-bill, in case Chicago Mill claimed the refuge effort was moot because the birds had already vanished.

Pough spent long hours in the field nearly every day from December 4, 1943, to January 19, 1944. On foot and on horseback, he searched for Ivory-bills and evaluated what habitat remained. "The racket from the 'cats' making it hard to listen for calls," he wrote to Baker. In another letter, he said, "So far no I-Bs. It is sickening to see what a waste a lumber company can make of what was a beautiful forest. Watched them cutting the last stand of the finest sweet gum on Monday. One log was 6 feet in diameter at the butt."

Neither Pough nor Baker used Audubon stationery in their correspondence while Pough was in Madison Parish. Baker had instructed Pough to avoid talking with the mill operator—a Mr. Ware, also the mayor of Tallulah—and with the logging supervisor, someone named Alexander. (A man named A. Carlson oversaw the operations from Truman, Arkansas.) Baker wanted Pough to keep quiet about Audubon's hopes and interests for the Singer Tract, for fear of further alienating Chicago Mill. Still fearing condemnation, the company could use Pough's reconnaissance as an excuse to accelerate logging. In his five-page, single-spaced instructions to Pough, Baker specifically recalled that Tanner had shared his logging recommendations with Chicago Mill—only to have the company cut more and cut faster.

Pough found that support in the parish for saving the forest seemed strong. Even Chicago Mill's local attorney, a man named Henry "Happy" Sevier, said he favored a national park or wildlife refuge. Pough stayed at Sevier's hunting camp in the woods, where he ate a diet of chitlins and squirrel and had to rely on the mail carrier to take him to Tallulah once a week.

What Pough didn't find, for days, was an Ivory-bill. Then, as an icy rain fell, he heard and sighted the only Ivory-bill left—a single female, the same one that Baker himself had seen months before. The bird's roost tree (with six holes) stood in Section 61 of John's Bayou, near a logging railroad and much cutover land, which the bird would not fly across. For nearly 10 hours in the rain, Pough watched the bird feed in small, second-growth trees—but not in sweet gums and not in the more extensive Ivory-bill habitat still available in Mack Bayou. In other words, Pough reasoned, this bird had become so faithful to this locale she would not leave it, even if that meant scaling bark from recently dead second growth instead of utilizing old-growth trees. Pough wrote to Baker, "Can't help feeling that mystery of I-bs disappearance is not as simple as Jim's report might lead one to believe. I wonder if the bird's psychological requirements aren't at bottom of matter rather than it being just a matter of food." Pough sent a telegram to John Baker on January 10, 1944, announcing his sighting, and, in a separate letter, declared, "I really fear the area will be cut any day."

Wildlife artist Don Eckelberry heard the staggering news and, in April 1944, headed to the Singer Tract to find the last Ivory-bill. He met a well-known local woodsman and parish game warden, Jesse Laird, who recently had supplied horses to James Tanner and who now drove the painter to John's Bayou. The Tensas River had flooded, so Eckelberry had difficulty moving across the wet ground jagged with stumps and slash. But, late one day, with dusk approaching, he heard—then saw—the lone female Ivory-bill. His were the last certain and authenticated sightings of the species in the United States. He spent two weeks following and painting the bird, joined once for a few hours by two young boys, self-described "swamp rats." (So impressed were the youngsters by the experience that one of them recently called Eckelberry, out of the blue, to reminisce.)

Eckelberry not only saw an Ivory-bill, he witnessed the logging. He rode a gas-powered car along the rail lines crisscrossing the Singer Tract and arrived in the ancient forest that U.S. Army guards derisively called "Little Guadalcanal." The guards kept watch over the German P.O.W.s who, of course, had no choice but to cut down the trees, despite being "incredulous at the waste—only the best wood taken, the rest left in wreckage." Eckelberry wrote that he "watched one tree come screaming down and cared to see no more . . . " The P.O.W.s loaded the logs onto trains that rolled into Tallulah where the sawmill was located.

World War II veterans of the British Army who sipped their tea in barracks or at the front—perhaps in distant North Africa—can thank the trees of the Singer Tract. "The Tallulah Plant was so busy making . . .

chests for supplying the English army with its tea that they had a regular production line which ended in 3 box cars sitting side by side on the railroad siding tracks," according to former Chicago Mill and Lumber Company official John R. Shipley. (One of Pough's 1944 letters also notes that sweet gums were logged to make a plywood for gasoline tanks in fighter planes.)

From time to time, the sawmill blades—fashioned like a belt—would catch on detritus from another war. As the sawmill dissected old-growth timber for British tea crates, the blades dislodged Minié-balls that had been fired by Confederate and Union troops.

The cutting sounds of a sawmill replaced, among other things, the sound of a flying Ivory-bill, whose beating wings James Tanner had once described as "wooden."

So the species is gone. Or is it?

While the evidence for the Ivory-bill's extinction is imposing, some birders and ornithologists cling to the prospect that these birds might yet be rediscovered. Or already have been. Sightings continue to this day, though they are never authenticated by follow-up observations, videos, audio recordings or photographs—so far as the general public knows. Ornithologist Jerome Jackson has said, "The bird still haunts scientists and bird watchers." This is true, but *is* the Ivory-bill alive? Or are some of us seeing ghosts?

Any chance that the Ivory-bill persists in the United States depends on the species being far more wide-ranging in foraging than James Tanner's work suggests. Indeed, both John Baker and Richard Pough privately expressed doubts about some of Tanner's conclusions, though not about his field observations. Pough noted that Nuttall's oak and sweet gum had been dying off throughout the South since the mid-1920s. In his January 1944 report to National Audubon, Pough asked:

Isn't it possible that these two [tree] species provided the bulk of the birds' food [during Tanner's study] because at the time they were the only trees that were slowly dying and therefore suitable as habitat for the flat-headed borers on which Ivory-bills feed? Possibly long continued dependence on these two trees produced a local population conditioned to this type of feeding.

Pough suggested that Ivory-bills needed dying or newly dead trees—*regardless* of species or age. It didn't matter how old or young the trees were, it didn't matter what kind they were—if they were dying or newly dead, then an Ivory-bill could feed in them. If this is the case, then perhaps the vanished Ivory-bill isn't gone, but just in hiding. Nancy Tanner says that her husband would have agreed with this assertion and that his focus on old-growth forests was simply due to his belief that only in such habitat could one find enough dying or dead trees, whether old or young. Birders walking in swampy forests south of, say, the Ohio River should therefore rivet keen attention on a flash of black and white and red. *Could it have been?*

Of course it's possible to make almost anything into something else, in the mind. Both the Pileated and the much smaller Red-headed Woodpecker have been visually mistaken for the Lord God Bird. Sometimes birders who believe they've *heard* Ivory-billed Woodpeckers have been fooled; they can mistake the call of nuthatches for Ivory-bills. Sometimes birders can't tell if they're hearing the recorded calls and drums on the Cornell tape that everyone uses to try to lure living Ivory-bills into view. Other people may have been confused by the sounds of a different creature altogether. A biologist at the Westborough Field Headquarters of the Massachusetts Division of Fisheries and Wildlife recently told me of his involvement many years ago in a search for Ivory-bills on South Carolina's Santee River. He later realized that the calls some team members took for Ivory-bills were probably the declarations of a frog.

The stories, though, the stories. Too many to discount. Sources too credible to dismiss. And with Pough's questions about Tanner's conclusions—well, what's a birder to think? Sightings of Ivory-bills have never ceased, even after the Singer Tract disappeared. (The birds probably could—can?—live between 15 and 20 years.)

Reports by respected naturalists Whitney Eastman and Alexander Sprunt, Jr., prompted the National Audubon Society—in cooperation with a receptive landowner, the St. Joseph Lumber Company—to open an Ivory-billed Woodpecker refuge on October 2, 1950. Warden Muriel Kelso patrolled the 1,300-acre Chipola Wildlife Sanctuary in Florida until the refuge closed on May 15, 1952. Nature writer Michael Harwood says Ivory-bills "visited their refuge once or twice a week . . . [though] after March 1951 they weren't seen or heard again." Sprunt claimed credible sightings in Florida's Apalachicola Swamp in early March 1952 and on July 10, 1952, in Wakulla County, east of the Apalachicola River.

Writing in 1987, the naturalist and author John K. Terres finally gave up his secret 1955 Ivory-bill sighting and related a story tense with awe and disbelief. On April 9, driving north on Florida Highway 19, near Homosassa Springs, with his wife, Marion, he saw "two large black and white birds" flying about 20 feet above him.

"What I saw I shall remember the rest of my life—*ivorywhite bills*, first, then the *broad area of white across both the forward and rear parts of their wings*. They were not pileated woodpeckers!

" 'Ivorybills!' I yelled."

John Dennis, who had taken authentic photos of Ivory-bills in Cuba in 1948, claimed to have seen a lone Kint in 1966, in the Big Thicket country of Texas. He said he saw the bird fly, then perch on a stump. Few believed him. After hunting in the Big Thicket for more than a week, James Tanner and Paul Sykes did not believe Dennis had seen an Ivory-bill. Dennis accused Tanner of "Ph.D. snobbism." Audio tapes Dennis recorded, with Ivory-bills calling, were attacked as fakes.

In 1968, someone took feathers from an alleged Florida Ivory-bill nest to the Florida Museum of Natural History in Gainesville. The Smithsonian Institution's Alexander Wetmore wrote a letter on April 27, 1968, to the museum—the letter is framed with the feathers—stating that a large white feather doubtless came from an Ivory-bill. But the feathers could be quite old and taken from a specimen in some collection. As yet, no one has conducted genetic tests to verify Wetmore's conclusion.

In 1971, respected ornithologist George Lowery, who had seen Singer Tract Ivory-bills in the 1930s, said he believed an amateur birder's photos of Ivory-bills—supposedly in the Atchafalaya region of Gulf Coastal Louisiana—were authentic. Lowery, curator of Louisiana State University's Natural History Museum, took umbrage at suggestions that the photos—never published—were frauds. Today, as well, photos or video-tapes would be met with similar doubt.

The supposed presence of Lord God Birds in South Carolina's Santee Swamp drew crowds of noisy birders, scientists and duck hunters in 1971. If there *were* any Ivory-bills, they might have flown off after hearing the hubbub.

In 1987, Jerome Jackson, one of the world's authorities on Ivory-bills, heard, with a student, what they considered to be a Lord God Bird. It called for many minutes in a forest north of Vicksburg, Mississippi, but they never saw the woodpecker. Eventually, Jackson writes, loggers cut that forest down. In the 1990s, reports emerged from Florida that Jackson says cannot be dismissed as mistaken visual identifications.

According to a student at Louisiana State University, Ivory-bills do survive—right now—somewhere along the Louisiana-Mississippi state line, in the Pearl River basin, where, in 1988, a biologist also says he saw an Ivory-bill. A senior student of Professor Vernon Wright's, at LSU's School of Forestry, Wildlife and Fisheries, recently filed a sighting report full of astonishing details. It may well turn out to be the last credible sighting report of Ivory-bills in America in this century. What is especially astonishing is that the observer claims to have seen and heard not one Ivory-bill, but two. A male and a female. At very close range. One negative feature of this report is its assertion that the underside wing pattern of black and white is "unmistakable." The Ivory-bill shows plain white on the trailing edge of the wings (and grayish plumage on the front edge), while the Pileated shows white only on the leading edge of the wings. Under poor conditions—and with adrenaline pumping—a person can confuse the two. Another point of consideration is the date of the alleged sighting: April Fools' Day, 1999.

After explaining that he was withholding precise location information from me, Wright says that he believes the report. Furthermore, he stresses that he had just lectured his students on the evils of falsifying data. Wright trusts the student and thinks "there are a few [Ivory-bills] left . . . [and] some of them may be nesting pairs."

On April 18, 1999, 16 observers went into the area of the alleged sighting and saw . . . nothing. "Several people reported hearing double taps on more than one occasion and some reported single high notes but no one felt comfortable with saying these were from Ivory-billed Woodpeckers," Wright explains. "Some of the sounds heard could have been from other individuals playing tapes."

This April Fools' Day report will remain—until independent and public verification—yet another tantalizing tale, a shimmering, lonely vision of the King of the Woodpeckers. Meanwhile, birders continue to survey the area, and officials may launch a formal woodpecker ecology study there.

Anyone—respected professor, enthusiastic student or backyard birder—can expect withering skepticism when claiming to have seen or heard this most elusive of specters. Perhaps other potentially authentic, late reports of Ivory-bills have been kept secret, in an effort to avoid the kind of intense grilling, even scorn, to which birders are subjected when claiming a hard-to-believe sighting. This is a well-known, even celebrated, part of science and of birding culture, a practice that developed most vigorously after shooting ceased. Without a bird in the hand, observers have to rely on extremely precise descriptions of plumage, behavior and field conditions. At meetings of area birders who have just finished a census— such as the traditional Christmas Bird Counts, which gather data on wintering populations across the country—census leaders cast baleful glares at the claimants, grimly interrogating them for contradictions or weaknesses. The ornithological equivalent of a grand jury deposition, this procedure often causes the post-count chili supper to roil in the bellies of potential witnesses.

Birders, conservation officials and ornithologists who know of credible Ivory-billed Woodpecker reports may have refused to offer them to the birding public, which would swarm like binoculared locusts into any region thought to have Ivory-bills. Some have argued for secrecy because property being used by Ivory-bills might be condemned by the Federal government, which could stoke animosity against the birds. Others believe that after appropriate measures have been taken—posting of extra wardens, for example—the return of Ivory-bills should be heralded. In either case, wildlife managers would face some difficult questions: How much appropriate habitat is available? Should more be purchased and/or protected? Should the birds be captured for safekeeping and, if both sexes exist, should captive breeding be attempted?

Perhaps no other bird species inspires the uncertainty that the Ivory-bill does—all of these questions and qualifications amid earnest wishes that it still live. But, according to ornithologist Lester Short, the Ivory-bill doesn't live; he believes the species is extinct, at least in America. So does The Nature Conservancy. A U.S. Fish and Wildlife Service biologist says the only reason the Ivory-bill has not officially been listed as extinct is administrative: It would take too much time and paperwork. Then he adds that the agency remains reluctant to declare extinct any species that has been sighted in the past few years.

Sightings of Ivory-bills in Cuba's remnant pine forests in 1986 and 1988 prompted the government there to take swift action, setting aside mountainous timberlands as a refuge and altering a highway route away from where the Ivory-bills were seen. A 1995 study declared, however, that the Cuban Ivory-bills were "almost certainly extinct"—victims of farming, logging, grazing and fire suppression, with the latter serving to deprive Ivory-bills of grub-infested, fire-killed pines.

But one prominent American ornithologist and a Florida biologist both confirm that they have heard about highly credible sighting reports from Cuba in 1999. Don Eckelberry says, "This is being kept quiet. I think people are afraid the place will be invaded by birdwatchers." The Cuban scientist whose team reportedly has seen the birds did not respond to my questions. May we take silence for affirmation? Some will.

The Lord God Bird may *not* be extinct, though it has vanished from the gaze of all but a few who have claimed to see real, living, breathing Ivory-bills. For the rest of us, uncertain and expectant, Ivory-billed Woodpeckers remain rumor, specter and desire.

For the rest of James Tanner's career, Ivory-bills flew and called in memory only. He had not returned to the old Singer Tract since his last visit in 1941.

Kelby Ouchley's phone call changed that. Ouchley, one of the first managers of the Tensas River National Wildlife Refuge, asked Tanner to come back to Madison Parish and tour the place that had once been the Singer Tract. About 40 years too late for Ivory-bills, the refuge had been established in 1980 and encompassed much of the woodpecker's former land.

Tanner agreed to visit and arrived on March 18, 1986. Ouchley and biologist Wylie Barrow both remember that Tanner was impressed by the extent of the refuge—50,000 acres at that time—and optimistic about the effort to restore bottomland forest there.

"Like everyone else's memory," Tanner had once written, "mine forgets annoying and unpleasant things and remembers the pleasing; the mosquitoes go but the bird song remains." But standing at Little Bear Lake, remembering the lay of the land he had not seen in decades, Tanner recalled the unpleasant things. "His birds were gone," Ouchley says. "I could see he was very pensive . . ."

Tanner had arrived in the midst of another controversy with an old nemesis—Chicago Mill and Lumber. Though the refuge was in place, the company was still logging. "The day he came we could see some of the old nesting areas were just cleared [again]," Ouchley tells me. "It was a pretty grim situation . . . midnight phone calls telling us the 'dozers had moved in and were clearing. There were [court] injunctions."

Today the refuge continues to restore a forest of locust, sweet gum, cypress, maple, ash and oaks—white, red, willow and overcup. The refuge shelters alligators, black bears, waterfowl and other birds, including the densest concentration of Barred Owls known anywhere. Farmers, Sierra Club members, teachers, politicians, state employees—all spent years lobbying to get the refuge established.

More than a half-century after he first arrived in the Tensas River watershed, James Tanner left that land for the last time on March 20, 1986.

He returned home to Knoxville where he had worked for years as a professor at the University of Tennessee and where Ivory-bills in the form of a print reminded him of his wilderness experiences. The print, hanging in his home, had been a retirement gift to Tanner from his faculty. The day that he and Nancy took the print to a gallery for reframing, the artist, unbeknownst to the Tanners, was there on business. Though Tanner would not have said anything critical, the artist asked him how he liked the print—so the scientist felt compelled to point out that the eyes were far too pale. Wielding a marker, the artist, nonplussed, quickly dabbed the correct yellow hue into the irises. Until his death on January 21, 1991, Tanner could watch the eyes, staring and precisely yellow, of his Lord God Birds.

Remind me again—together we

trace our strange journey . . .

—William Stafford, "Our Story"

I stand in Mack Bayou. I stand on ground that was the Singer Tract. It is September, the autumnal equinox, the time of molting. Sixty years after a lone male Ivory-bill lived here, 60 years after the logging began, I have come to see the forest returning, to pay some kind of respect, to learn the textures of this poignancy and resolution. All morning I've walked on a wide trail—an old logging road, I'd guess—in shade and sun. All morning I've noted bear scat and watched, amazed, the hundreds of hackberry butterflies attending me like barnstorming sprites, landing on my sweaty shirtsleeves to feed.

If I walk in the light and shadow of a forest growing back, if I walk among the memories of trees and birds, if I walk into grief, like this thick air, I want to walk not away but through. To make a crossing, if not to courage then to duty or vigilance or whatever one might name the desire (admittedly selfish) to live with as little regret as possible in relation to the world—and to gesture some honor, some witness, toward these other lives, from whom we have taken so much, even themselves.

I stand at a bend in the trail, where, if I weren't afraid of startling bears, I'd bushwhack through green-fanned palmettos and prickly vines and try to stand as close as possible to where one of the last Ivory-bills' roost trees had stood in 1938. In his unpublished essay, "A Postscript on Ivory-bills," Tanner recalls this individual, nicknamed Mack Bayou Pete:

> [He] was hard to find, and once found was soon lost. Restless as a blue jay, Pete would dig out a few borers, pound and call loudly, then wing through the tree tops to some distant part of his domain, leaving the earthbound watcher to plod over (and through) the forest floor in the same general direction with eyes scanning the trees and ears stretched wide to hear a distant *kent*.

I drop my pack, set my binoculars down and remove a tape recorder.

I face northwest, toward the ground where, about a half-mile away, Pete's roost tree stood. Beside a palmetto and a Nuttall's oak sapling, I turn on the tape. For me, Ivory-billed Woodpeckers exist as words and as bodies of air, their calls captured (as we say) first on films and transferred later to a tape cassette I ordered from Cornell University's Library of Natural Sounds: LNS number 06784, "Calls and Drums" of the Ivory-billed Woodpecker, *Campephilus principalis*. In a portable cassette player, I carry sine waves of a species, a moment of its life, traces of a past.

"Ivory-billed Woodpeckers, Cornell Catalog, Cut 1," announces Paul Kellogg's voice with Hollywood flair. From the tape player comes a plaintive, truncated, high-pitched and honking sound—the nasal *kent* of the Ivory-bill. And in the background of the tape, the trill of a Red-bellied Woodpecker. The call of an Ivory-bill has the foreshortened quality of a car alarm—beep—as it's deactivated by its owner and therefore, for me, lacks the majesty of the bird's appearance. On the tape, two Ivory-bills call to each other. These are the nesting birds that the Cornell expedition recorded in John's Bayou in 1935. *Yent, yent, yent* I hear. Then the loud double rap of a bill striking wood.

I hold the tape player out as if I were holding a badge or a tiny shield until the sounds of the Ivory-bill end abruptly. I think of how Tanner described Mack Bayou Pete's voice as " 'his head in a bucket' sound." How each day Pete flew a couple of miles from his roost tree across grub-poor woods to feed mostly to the east and north. I stare at the bright sky, the shadowed forest and wish I had seen that bird. A juvenile with more white on his primaries than usual, Pete stayed another year or so, then disappeared.

Some psychologists, scientists and activists have written about the parallels between familial grieving and ecological grieving, comparing the loss of a treasured place or species to the loss of a loved one. If the comparison rings true, and it does for me, then we must find ways to grieve well. We must confront loss rather than deny it and, in doing so, nurture the energies to cope with the difficulties of loving a world we have systematically diminished.

After my brief ceremony, I listen to lives giving song—the persistent White-eyed Vireos—and, eventually, drive back across the bridge over John's Bayou. Though I am tired from my hike, I get out of my truck and stand beside John's Bayou with its narrow band of trees snaking beside the stream. On either side of the few trees flanking the bayou are vast farm fields. I close my eyes and see a jungle instead.

The day after my bayou sojourn, I ride along in a pickup with R. T. Williams, a retired state game warden and biological technician for the Tensas River National Wildlife Refuge. The unofficial historian of this place, R. T. speaks dramatically, with an accent as thick as his arms. Turning off Road 884—"Sharkey Pl'nation Road"—R. T. steers the truck along back roads and trails, pointing out the locations of Chicago Mill and Lumber land, as well as old railroad spurs and Indian mounds destroyed by loggers.

R. T. soon pulls into a sunlit clearing that serves as a hunter's check station. The sign by Sharkey Plantation Road reads, "Report Game Here." An inverted L-shaped game pole stands by a tree in the clearing. "The old Crocklin camp," R. T. says. "Then Kuhn had it for a while. James Tanner stayed here . . ." We get out, and I bend briefly like a supplicant before the rusting parts of a 1920s- or 1930s-era car.

Later, alone, on my way out of the refuge, I see that the door to what was Jesse Laird's house is still open. Here James Tanner borrowed a horse when he would trek into the Singer Tract. Today the abandoned house on Sharkey Pl'nation Road hosts a rancid mattress, an old fork and mosquitoes. Moss grows on the roof.

Back in Tallulah, I drive past Popeye's Chicken, site of the old Post Inn Hotel, where the 1935 expedition likely stayed, then head toward the Tallulah Correctional Center for Youth, with its chain-link-and-razor-wire fence and inmates in orange jumpsuits holding weed whippers. The

*The old Chicago Mill and Lumber Company sawmill in Tallulah,
near the Singer Tract.*

old mill that processed the Singer Tract lumber sits on the prison property. But a guard won't allow me to look at the mill even from the prison
parking lot.

So, a couple of minutes later, in town, at the corner of West End and
Allen, I gaze back toward the ruins. Made of red brick, and partially collapsed, the facility shows the bones of metal girders. The large lot where
men once stacked ancient trees grows over with grass and weeds. Vines
climb the water tower. The smokestack reveals exposed brick in patches,
where a layer of concrete has fallen away. A long chute rusts from the
metal bin on top of the brick building all the way down to the chute's conclusion in the weeds.

To leave Tallulah, I drive out the way I came, through Waverly—a few
mobile homes and a cotton field—then, at Tendal, I cross again the Tensas River, which is just a muddy creek right now. Roadside ditches release
clouds of sulphur butterflies that hover and fly above and across the road

and the fields, cotton and fallow, into whose earth the metal legs of power poles thrust, their lines crackling to the four directions.

When I return home I'll put a videotape in the player and watch again the film the 1935 expedition made in the Singer Tract, a film of short silent scenes: Paul Kellogg unloading gear, James Tanner marching through mud—or trying to march—the men walking and riding into the wet forest, Allen (I think) gesturing toward a tree, then the jump cut—a moving, living Ivory-bill at the nest hole. The huge bird—the female, I think—jerks her head in and out of the hole, as she clings to the tree in profile against the forest sky, a few branches blurry in the background. Then she looks with bright eyes over her right shoulder, as if seeing something. The film concludes with Arthur Allen at the binoculars in camp and James Tanner at the sound mirror. In all its archival reality, the film is both distant and intimate, like so much that has to do with the Ivory-bill, including this final story.

It comes from Thomas Barbour's book *The Birds of Cuba*, and concerns an injured female Ivory-bill, trying to cope with, Barbour wrote, an "... upper mandible [that was] monstrously long ... The long and curled bill had grown so that the bird could not secure its own food." Unable to forage, the female Ivory-bill would have died had she not received food, assiduously, from her attentive mate.

If we want things to last, we must take such care.

The Heath Hen

Overleaf: Alexander Wilson's "Pinnated Grous."

The Fire Birds

MY CHRONICLE OF THE HEATH HEN—A BIRD OF THE EAST—BEGINS IN A
Kansas parking lot. Standing in the predawn April chill, I waited for a van
to arrive that would take me to a nearby prairie, where brown-and-white
striped Greater Prairie-Chickens gather each spring, as they have for
thousands of years. These birds—the still-living cousins of the now-
extinct Heath Hen—enact on their grassy hilltops a series of mesmeriz-
ing and ludicrous courtship displays, rituals that provoke in watchers
feelings of bemusement and awe. Lit by tall streetlamps, I zipped up my
coat, shoved my gloved hands into pockets and considered: Whenever I
see a Greater Prairie-Chicken I see a bird whose appearance and behavior is so close to the vanished Heath Hen it's as if I am looking into the
past.

These two races of one prairie-chicken species, *Tympanuchus cupido*—the eastern race of Heath Hens, the midwestern and prairie race of
Greater Prairie-Chickens—at some time indeterminate, and for reasons
unclear, became separated and therefore lived apart in their respective
geographies. Historically, Heath Hens frequented dry, brushy habitat
characterized by low trees—stunted scrub-oak and blueberry barrens—
as well as grassy clearings and meadows, perhaps as far south as the Car-
olinas and as far north as Maine. Long Island, New York, New Jersey, as
well as Pennsylvania, Connecticut and Massachusetts, constituted the
Heath Hen's stronghold. The Greater Prairie-Chicken, however, pros-
pered in tallgrass prairie dotted with oak woodlands wherever such habi-

tat existed, as far east as Ohio and Kentucky, south to Arkansas, west to Kansas, Nebraska, Colorado and the Dakotas and as far north and west as parts of Canada, after the spread there of small farms settled by pioneers.

Although the extinction of the Heath Hen meant the end of that race, it did not mean an end to the species. About 400,000 square miles of native tallgrass once grew in North America. Less than one percent remains. What persists, in conjunction with acreage of exotic grasses, is enough to support somewhat sizable Greater Prairie-Chicken populations in such places as South Dakota, Kansas, Nebraska and Oklahoma. Much smaller—and often locally threatened or endangered—populations occur in Illinois, Missouri, Minnesota and Wisconsin. In Colorado, the population is on the rise.

The morning I woke at 3:30 for my prairie outing, I drank coffee and looked at photos and paintings of Heath Hens and Greater Prairie-Chickens. (Naturalists called them both pinnated grouse—a more euphonious name than "chicken.") For all I could tell, I might have been looking at different snapshots of the same bird: chicken-sized, plump, thick-legged, short-tailed, with a plumage of narrow vertical white-and-brown barring not unlike the stripes of a zebra.

If the Dadaists had also been ornithological divinities—if Tristan Tzara had formed a committee to design grouse—they would have conjured *Tympanuchus cupido* from the found art of the animal kingdom and the human realm, using ready-made appendages, shapes and colors: the white rump of a white-tailed deer; the short, stiff, angular tail of a Ruddy Duck; wing feathers that extend like knives beyond the rump; flanged, dangling neck feathers called pinnae that courting males can lift like pointing fingers; the creamy dun-and-chestnut streaks of various grouse; and, of course, the male's featherless air sac on his neck, which he can swell up to the size of an orange and which is that color or perhaps more nearly the color of Tibetan Buddhist robes. Heath Hens and Greater Prairie-Chickens, more brothers and sisters, I suppose, than cousins—nearly identical twins, really—are avian *assemblages*, surreal chickens.

Even the scientific designation *Tympanuchus cupido* reflects this curious melding. The genus name refers to the male's bassy call: *boom,
boom, boom.* Timpani. When Linnaeus, the naturalist who constructed
this classification, examined a male's narrow neck feathers (which droop
down until the bird lifts them straight up like horns) he thought of a
god, he thought of Cupid's wings. Hence, *cupido.* A drumming love-god
chicken, surreal indeed.

Because of the similarities in appearance, I can distinguish Heath
Hens from Greater Prairie-Chickens only through history. So this other
metaphor may not be original, but it's accurate: Right now the Heath
Hen lives only as a doppelgänger to the Greater Prairie-Chicken, with
whom, in all its goofy glory, I was about to be reacquainted.

When the white van pulled up, with John Cavitt at the wheel, I
grabbed my binoculars and fanny pack stuffed with snacks, then climbed
into the overheated, beat-up Ford. Fellow members of the local Northern
Flint Hills Audubon Society—Sue Dwyer and Hoogy and Carol
Hoogheem—also nestled in. John, a doctoral student in ornithology at
Kansas State University, drove us out of town on empty streets past dark
storefronts. John explained that he would take us just outside of town to
a little-known entrance on the eastern side of the Konza Prairie Research
Natural Area—a Nature Conservancy reserve whose 8,600 acres of tallgrass provide scientists with long-term opportunities for an array of ecological studies and experiments. From there, we would turn onto a dirt
road that would lead us to the prairie-chicken blind, a narrow, wooden
shack moved occasionally from lek to lek in order to minimize disturbance to the birds. We'd arrive beside the lek—the display grounds of the
chickens—just before the birds gathered, in order not to spook them.
Once in the blind, we'd stay put and stay quiet, regardless of cold or blad-
ders. The birds would be in charge. We'd attune ourselves to their needs.
We'd leave only after the morning courtship show had concluded. (If we
wanted, at dusk, we could see a reprise, for the males and females gather
again in the evening to attend to the serious business of mate selection.)

I kept silent for a few minutes, looking out the dirty window at stars in the east: The Summer Triangle of Altair, Deneb and Vega burned there, familiar markers of seasonal change, sidereal return. I had been getting to know Sue, Hoogy and Carol better (all recent arrivals in Manhattan, Kansas) through our conservation volunteering and birding. As the broad road cuts of K-177 flashed by—exposed white rock, limestone layers, time's compression in stone—I felt this sense of community extend to the land itself, to the steep-sided, flat-topped Flint Hills and the lives dependent on this last vast expanse of tallgrass and then on to another country, the sky, where stars long gone had created the very atoms of this morning's world.

The Greater Prairie-Chicken is a signature species, as vital to tallgrass as bison; on another watershed of the Konza Prairie a herd of about 200 bison roams and grazes, a remnant mimicking of the measureless herds of decades past. Writes avian ecologist John Zimmerman, "As long as these birds' overture to the rising sun can be consistently heard upon an April morning, you can be assured that the true prairie remains alive and well, at least in that locale." The prairie-chicken seems even more vital when one considers what is missing from tallgrass—the Plains grizzly bears and wolves that once thrived here.

If the prairie-chicken lacks the mammalian heft and power of bison or bear, it compensates for that lack by its springtime exuberance, its captivating courtship that lends to cold April dawns an odd passion. N. S. Goss described it in his *History of the Birds of Kansas*:

In the fall the birds collect together, and remain in flocks until the warmth of spring quickens their blood, and awakes the passions of love; then, as with a view to fairness and the survival of the fittest, they select a smooth, open courtship ground . . . where the males assemble at the early dawn, to vie with each other in courage and pompous display, uttering at the same time their love call, a loud booming noise; as soon as this is heard by the hen birds desirous of mating, they quietly put in an appearance, squat upon the ground, apparently indifferent observers, until claimed by victorious rivals, which they gladly accept, and receive their caresses.

As the van pulled onto a dirt road, I sat up expectantly: In a few minutes courage and pomposity would display themselves, personified in feathers, hyperbolic as purple prose.

We soon piled out of the stuffy vehicle, put on daypacks and gloves and began our walk to the blind, which, still in the dark, we could not see. I looked over my shoulder more than once, hoping for a final glimpse of Comet Hale-Bopp, so I stayed last in the single-file line and wondered why those farther ahead kept kneeling.

Smiling, John held apart the barbed wires of a fence. "I wasn't sure why you were crawling," I said, easing through. "I couldn't see the fence. I thought maybe you were making us kiss the prairie before we get to the blind." Outlined by dawn light, the low shack faced level ground to the west—the lek or booming grounds, the stage for chicken passion. (Some biologists still prefer to call prairie-chicken leks by the old term of booming grounds, because leks are communal courtship display areas in general; many kinds of species use different kinds of leks.)

At 6:10 A.M. we were in and done adjusting parkas and gloves. Then gray hints of light brought gray premonitions of sound, then gray sounds themselves, two or three bass coos, like the notes of deranged Mourning Doves. A few minutes later a few dark shapes emerged in the grasses before us—male Greater Prairie-Chickens. More bass coos, almost continuous, like a droning, an undertow of sound. *Uoooo-oooo.* Next came mewling cackles, and one sounded as though it was directly in front of the blind. I closed my eyes.

I heard a slow, low *o-o-o* repeating and repeating, textured with pauses and, I thought, slight inflections. Naturalists once called the booming of the males "tooting," which connotes perhaps too high a pitch. Others have translated the call, somewhat loosely, to "old muldoon," which could be the name of a bar. Neither "tooting" nor "old muldoon" captures the hypnotic moaning quality of several birds booming and droning at once. The o-o-o of the Heath Hen and the Greater Prairie-Chicken has been compared to the sound of blowing air over the lip of an empty glass bottle as well as to the notes of faraway tugboats. One writer said of droning Heath Hens, "It sounded as if a lot of lonesome little night winds had taken to crying whoo-oo-o in ragtime, mingled with whistles of syncopated measure . . ."

The booming moans emanate when air from the bird's syrinx, the vocal organ, hits the skin of the inflating air sac puffing out on the neck. The air sac then amplifies the sound, which can carry a surprising distance. "On certain mornings . . . I could distinctly hear the tooting of the birds . . . more than a mile away," wrote Alfred Gross, who studied Heath Hens on Martha's Vineyard, the island off the Massachusetts coast. Gross found that the birds boomed most readily on still, cloudy days, but often stayed silent on very warm, clear mornings.

Imagine chanting monks. Imagine cackling monkeys. Imagine cold air and the feet of chickens rustling grasses. Nineteenth-century ornithologist Thomas Nuttall believed "that at a distance the sound [of Greater Prairie-Chickens booming] might be taken almost for the grunting of the Bison, or the loud croak of the bullfrog." In 1927, Edward Howe Forbush, the state ornithologist of Massachusetts, sought to distinguish the nearly identical sounds of the Heath Hen from those of the Greater Prairie-Chicken; Forbush said the Heath Hen's booming "is not so deep and resonant as is that produced by the prairie-chicken" and does not sound like bison or frogs at all. "It may be likened," he suggested, "to the soughing of the wind . . ."

In the day's first ample light, we in the blind could see the birds' soft brown bodies, the fences, a stock tank and a pair of frozen ponds. We heard the pitch change—a bit higher, I thought—oiue. And wohoo—perhaps a female (or hen) skulked nearby, but we couldn't see whether one elicited this peculiar outcry. The males (or cocks) purred with laughter. I could see one chunky shape then another and another. Rapid cackles—"petulant" said one naturalist—like a chitter, then a *whoop*. John confirmed what I had suspected—that males "whoop" only when they think females are present. (Sometimes males "act" like females, causing mistaken *whoops*—a rather apt word, in such a case.) And the cackles are more like male conversations, John explained. Guy talk. When two males came too close to one another, the cackles increased riotously. (Gross found that when a male Heath Hen intruded on another who was booming, the offended cock opened his beak and squawked instead of boomed.) The lek—a flat hilltop with tawny shortgrasses and patches of dirt—afforded no more room than a suburban house lot, so the males found themselves increasingly rankled.

Soon, 11 males had gathered, charging each other with quick, pattering steps. Visualize fencers running at one another with baby steps. Sometimes the charges ended with the two combatants facing off, often in a squatting position, each looking across some invisible boundary with their hard dark eyes. Sometimes they charged and then, studiously nonchalant, they'd just decide, well, perhaps it's better to wander over here and peck at the dirt. Impressively, males jumped high into the air—two proud gymnastic challengers—with feet splayed out and an occasional feather flying off. Rarely do these encounters cause real injury. Greater Prairie-Chickens are the professional wrestlers of the bird realm. They prefer choreography to harm. Some of the cocks simply ran at targets unseen by human eyes. Perhaps the air itself offended these chickens' dispositions.

Over and over, for many exhilarating, exhausting minutes, the males erected their pinnae in very spiffy black *V*'s—lusty little devil horns—then bowed forward so their bellies paralleled the earth. They spread their tails, pointed their wings to the ground, bobbed their necks with a quick jerk and inflated their orange air sacs, swelling like party favors. The birds charged and squatted hard down, again. They jumped, again, turning full circles or half circles in the air, then landed, nonplussed or affronted. Some males turned partial circles while stamping their feet on the ground, nature's most perfect tantrum or celebration or both. It seemed that no vanity of posture, no absurdity of motion, could finally uncoil this profuse, enraged desire.

Sue Dwyer leaned over and whispered, "*This* is the origin of the end zone dance."

Amid this serious carnival, we noticed a single male that kept apart on the north side of the lek and another loner that stayed on the periphery near the stock pond. John Cavitt explained that these two cocks on the outskirts probably were young males. The biggest losers, in short. Most males go unrewarded for their courtship efforts because the older, dominant ones get most of the breeding opportunities with the hens. That is one of the points of gathering on the leks. Even so, less dominant male prairie-chickens do their best to vie for attention and they even, sometimes, manage copulations with females moving from lek to lek. We saw no copulations that morning on or off the lek, though if we had we would have seen other males scream in protest and probably attempt to dislodge the lucky fellow from his mate.

When a female arrived on the fence, then flew to a perch on a low sumac branch curved and glazed with ice, the single-minded severity of *Tympanuchus* hormones burst forth in a kind of Dionysian spectacle: prodigious stamping, cackling, booming, cooing and general circling about with proud wings dragging the ground. It was pathetic—and wonderful. The female watched, unimpressed, flew to the ground and searched for breakfast. Meanwhile, the bird I had dubbed the Stock Pond Loner inflated orange air sacs and erected his neck feathers—*V*! I'm here! I'm, I'm a prairie-chicken! No hen paid him attention, though we humans offered sympathy.

"Better lek next time," someone said.

Slowly, as we watched, we thought we could discern a pecking order not only among the males but also among the females. One particular hen seemed to elicit far more squealing from the males than did other females, but she remained shy and acted disinterested. In fact, all the hens watched the males carefully, as they considered with whom to mate. The females, though drabber in plumage, possessed a greater sobriety than did the males. Perhaps the hens were wary of the nesting and parenting duties ahead, with which the males have no part. The females will select shallow ground nests in tallgrass near thicker cover, which provides some security from predators such as foxes and hawks. After having mated with one or more males, the females typically lay about a dozen eggs, which they incubate for just over three weeks; the precocial chicks can fly about a week and a half after hatching.

Heath Hens were ground-nesting birds as well, locating among leaves, grasses and ferns beneath scrub oaks and berry bushes. Not even Alfred Gross, who wrote the definitive history of the Heath Hen, ever found one of these well-camouflaged nests. Like the Greater Prairie-Chicken, the Heath Hen laid yellow-brown eggs sometimes marked with dark splotches. Hatching out in June or July after three and a half weeks of incubation, the Heath Hen's precocial broods averaged between 5 and 10, at least on Martha's Vineyard.

"A female was just stamping her feet," Hoogy commented to the rest of us, and another female flew in to join the lek. We tried to count—four or five, even six hens now? We craned to look and quantify. In the very act of watching and trying to know there can arrive a kind of redemption and renewal. A forgiveness moment. The sun suddenly spilled light, as three females, looking long-necked, pecked in the shadow of the blind. Ice-glazed sumac branches glittered in the cold spring sun. Ancient tendencies marked the day with such immediacy.

After a Northern Harrier flew swiftly overhead and frightened the prairie-chickens—who seemed to evaporate—we left the blind and headed for home. While walking to the van, I stopped to watch a Red-tailed Hawk catch the morning's first thermal, an invisible swirling rising; beneath my feet, beneath the rocky soil, lived the tendriled rhizomes of prairie grasses. Some of these rhizomes could be several hundreds of thousands of years old. The world wending . . . thermal, hawk, rhizome, prairie.

John Cavitt drove us out on a different gravel road, which cut between several of Konza's many, intensively studied watersheds. To the right were tawny stalks, standing dead tallgrasses; to the left nothing but black char dotted with countless, flat limestone rocks. Such blackened earth looked like disaster the first time I saw a range fire's aftermath here, years ago. Now I know better.

Fire created the Heath Hens' home, just as it creates the Greater Prairie-Chickens' home. Set by lightning strikes, Native Americans and, on Martha's Vineyard, blueberry farmers, fire kept forests from establishing themselves, thus generating the brushy scrub-oak and berry-bush barrens Heath Hens needed for foraging and nesting, as well as the meadows needed for booming grounds or leks. (Scrub-oak roots can withstand brutally hot flames.)

On the Kansas prairie, dead tallgrass stalks can stand for many years, shielding nearly all the sun's light from the earth, slowing the growth of soil fungi, which need more warmth; ultimately, this lowers the amount of nitrates available to microbes and plants, therefore reducing plant productivity.

Now set deliberately by ranchers in the Flint Hills, fires clear off the old vegetation and produce ideal conditions for the warm-season grasses that here predominate: big and little bluestem, Indian grass and switchgrass. Bison especially love to graze on big bluestem and on the tasty green shoots of new growth that appear just days after a scorching prairie fire. Cattle largely have replaced bison as the primary grazing ungulates, but the role of fire remains central to the prairies.

One can have, though, too much of a good thing. Flint Hills cattle ranchers have increased the frequency, intensity and expanse of spring "controlled burns"; following the fires and regrowth of grass, more cattle than ever graze the prairie. Because Greater Prairie-Chickens prefer to live in tallgrass habitat that has been burned about once every three years and tend to nest in prairie only slightly browsed by cattle, these more frequent burns greatly diminish nesting cover in late spring and early summer.

We tend to think of bright birds, red and orange birds—orioles and tanagers, for example—as "fire birds." But those who study and love brown wild chickens know these are the true fire birds. Whenever I watch a prairie fire, I see in the leap and skitter of flames the energy and fervor of dancing prairie-chickens.

If Greater Prairie-Chickens begin to show long-term population declines as a result of too much fire and too much grazing, yet another clash between environmentalists and the agricultural community will result. This has happened already with the Lesser Prairie-Chicken—a paler species of the western shortgrass prairie—whose habitat has been so fragmented and whose numbers have so plummeted that one environmental group has petitioned the federal government to list the Lesser as an endangered species. The U.S. Fish and Wildlife Service so far refuses. Meanwhile, the Attwater's Prairie-Chicken—the southern race of the Greater Prairie-Chicken—barely holds on in Texas. This highly endangered race numbers only in the dozens; its coastal prairie habitat has been decimated by development and agriculture.

As stand-ins, as ghostly doubles for moments long gone, the morning's events on Konza Prairie gained the power of the totemic in my eyes, as if history had become animate. Every action on the lek brought to mind Heath Hens, the details of their lost lives, and the attachments I felt to them. History's rhizomes may not be as tough as those of switchgrass, but they run beneath the surface of our lives in places. They hold the soil, too, if we choose. They remind us of connections that span a continent and the years. The last holdout of the Heath Hen, for example, was an ocean island. And this place in which I live, the Flint Hills of eastern Kansas, is the stronghold of the Greater Prairie-Chicken only because an ancient sea once covered this part of North America.

The bodies of hordes of small Permian sea creatures eventually became limestone while, geologists speculate, the siliceous shells of these or other creatures hardened into flint or "chert." Because these flint nodules occur so thickly among the limestone shelves of the steep-sided Flint Hills, settlers could not plow this prairie. They tried and found flint so resilient it literally raised sparks from blades. In battles between flint and steel (at least before the development of the internal combustion engine), steel lost. So it was that an accident of Permian geology saved this land from a complete erasure of North American prairie that has proved so devastating elsewhere. (Lands broken out later for crops often lost money, and so were returned to grass pasture.) The Flint Hills—America's last great unplowed bastion of tallgrass—repeat grassy, flat-topped ridge after grassy, flat-topped ridge from the Oklahoma border nearly to Nebraska. This region ranges from a few miles to some 50 miles wide and comprises about 7 million acres or 11,000 square miles, 13 percent of the total land area of Kansas. Private livestock producers own nearly all of those acres.

The conquistadors and settlers compared this country's wind-waved prairie to the actual sea, the ocean crossed to reach the New World. We've so often remarked in prose, poem and song on "the ocean of grass," on "the amber waves"—whether of bluestem or wheat—that tawny, stalky waves endure as one of Kansas's most famous clichés, along with its supposed pancake flatness and Dorothy's too-oft-repeated observation about having left Kansas behind. Born from ocean, the Flint Hills—in their undulation, the lift and swell of rock, dirt and grass—echo Atlantic swells which encircle the island that was the final home to the Heath Hen.

In 1602, Bartholomew Gosnold crossed the Atlantic and arrived at an island off the New World coast; it's said that he called the place Martha's Vineyard in honor of his daughter and in recognition of numerous vines of wild grapes he found there. Gosnold may have encountered Heath Hens during his foray on Martha's Vineyard or another nearby island.

His crew included a friend of William Shakespeare—a man named John Brereton—who probably told Shakespeare tales of the islands. Some have speculated that the bard based *The Tempest* on such tales, which must have included descriptions of unearthly sounds, the booming of Heath Hens and their cackles, the moanings of a deformed spirit, the voice of Caliban, and perhaps Ariel—or both.

As Gosnold and others explored this new land, the Heath Hens continued their ancient, encoded ways. Waking before or at dawn in blueberry barrens in Maine and in scrub-oak plains in Massachusetts—starting the day with a shiver and shake of feathers up and down the eastern seaboard—Heath Hens set out for the important work of survival: foraging for food.

In the spring, Heath Hens favored grasses, clover, sorrel and other new growth. In summer, they would eat insects and berries, including bayberry and blueberry. Heath Hens especially liked the leaves and fruit of partridge berry, so much so the plant came to be called Heath Hen plum. I looked in vain for partridge berry when I visited Martha's Vineyard and eventually found it along a New Hampshire nature trail, with a little sign identifying the plant. The birds added various seeds to that diet in the fall, and, come winter, the birds also consumed acorns—what some called the bread of Heath Hens.

An opportunistic generalist, eating a wide range of plants and insects, Heath Hens seeking strawberries even wandered into the gardens of farmers and townsfolk. The tiny beaks snipped off the red fruits meant for desserts or salads, but in return Heath Hens sometimes helped gardeners by eating potato bugs.

After morning feeding—feasting even, for Heath Hens could be voracious eaters—the birds would bathe in dust and dirt to rid themselves of parasites, then would find shade and loaf through the afternoon. Hunger sent them out again, later in the day, and they sometimes ate well past dusk. Occasionally the birds would slide open beaks along leaves or grass blades in order to sip dew or raindrops, though most of their needed water came from the food they consumed. Nights, the Heath Hens roosted on the ground in scattered groups.

The lives of Heath Hens were also governed by the rhythms of reproduction. In the spring—especially in April—the birds would seek out the wide fields of cropped grass to gather for booming, for courtship displays. Summer the females devoted to raising the next generation, each night bringing peeping chicks under their wings for sleep. Fall offered another opportunity to gather at the leks, though cockfights and booming always proved less vigorous than in the spring and did not trigger mating. Winter prompted flocking. There was safety in numbers and probably greater success in locating acorns and other foods.

So went the days and years of the Heath Hen, for generations, for centuries. The bird's abundance had come to exist precisely because of its fidelity to daily and seasonal behaviors honed by natural selection.

But the Heath Hen suffered catastrophe because the new Americans took advantage of those behaviors. Wending back to stories of the first significant American settlements, we find that the Heath Hen provided not just possible poetic inspiration; it provided meat. "Heathcocke . . . [are] common: he that is husband and will be stirring betime, may kill halfe dozen in a morning . . . their price is four pence a piece," wrote William Wood in 1635 in *New England's Prospect*.

Some writers insist that the Pilgrims survived—especially through their first New World winters—because of the abundance of Heath Hens. According to Thomas Nuttall, the birds "were so common on the ancient bushy site of the city of Boston, that laboring people or servants stipulated with their employers not to have the *Heath-Hen* brought to table oftener than a few times in the week!"

Wrote J. H. Studer in his *Popular Ornithology*, "The flesh is dark, having a gamy flavor, and, where not too common, is considered a great treat." Another observer claimed, however, that "as for eating them, such a thing was hardly dreamed of, the negroes themselves preferring the coarsest food . . ." Certainly after midautumn and through the winter, the Heath Hen's diet of scrub-oak acorns imparted bitterness to the bird's flesh, a fact that may have contributed to its poor reputation among some. It tasted best when cooked just hours after killing and when consumed in the early fall, before the birds had shifted to eating acorns.

While there may have been a preferred season for bringing the tastiest Heath Hens to the table, on any day of the year there were gunners hunting the birds. Spring mornings especially suited these men whose interest was killing for sport or profit. On such mornings the Heath Hens gathered on their leks, making it easy to dispatch them by the dozens, even by the hundreds, as the gunners fired from what Nuttall called "ambuscades of bushes." Nineteenth-century New York naturalist Dr. Samuel L. Mitchill paints a vivid, disturbing scene:

> The destroyers construct for themselves lurking-holes made of pine branches, called *bough-houses*, within a few yards of the parade, and hither they repair with their fowling-pieces, in the latter part of the night, and wait the appearance of the birds. Waiting the moment when two are proudly eyeing each other, or engaged in battle, or when a greater number can be seen in a range, they pour on them a destructive charge of shot.

Thus the gunners drove Heath Hens from many of their ancestral booming grounds.

Before colonists and guns appeared, Native Americans would sometimes spread ashes on the leks, so that the strutting, stamping, dancing cocks raised up a blinding cloud. With the birds unable to see, men set upon them with sticks, clubbing the birds to death. (Native American hunting—clever as it was—did not seriously harm the Heath Hen's population, as demonstrated by the vast numbers of this bird when Europeans arrived.)

Not only easy to kill on its scratching grounds or leks, Heath Hens made obvious targets when they flew up into low trees, which they did frequently, much to the pleasure of the gunners. Given the straightness of their flight, Heath Hens made easy targets in the air as well. Several nineteenth-century sportsmen wrote that the Heath Hen provided so little challenge that the bird was best left to gunners shooting for public markets and to youngsters honing their hunting skills.

On the ground, a flock being pressed by hounds would run through grasses, the birds repeatedly crying *coo-coo-coo* and some squatting down as if the danger would pass them by. Usually, no more than two birds at a time would pop up into the air, enabling hunters to shoot many birds over a few minutes as the plump targets flew from grass to sky at steady intervals. If all that weren't enough to pummel the Heath Hen's numbers, the birds were also baited and trapped.

So the Heath Hen began to fall to a killing frenzy that provided tremendous numbers of birds to colonies and settlements, then to the growing markets of eastern cities, where prices fluctuated wildly according to supply and interest. In 1821, two Long Island Heath Hens fetched a hefty $5, with the price doubling within five years. Even if some people did not care for the taste of Heath Hens, they could be easily obtained. The birds filled up the wooden market stalls in Boston, New York and elsewhere. When local populations of Heath Hens dwindled, commercial hunters turned to the western Greater Prairie-Chickens to fill the void, so deluging the market that by 1861 a pair typically cost just 50 cents.

The massacre of Heath Hens caught the attention of the New York State Legislature as early as 1708, when some public officials sought to protect the birds on the Long Island plains. In 1791, that assembly considered a bill that some representatives had understood to offer protection to *heathens*. After the misunderstanding cleared up, the bill passed.

These and other state attempts to protect Heath Hens proved useless. Consider Massachusetts, which in 1831 declared spring off-limits for hunting the birds. Even if enforcement existed—which it didn't—the fine amounted to only $2, typically less than the market value of two Heath

Hens. Massachusetts continued, off and on, to make utterly ineffectual attempts to preserve the bird. The state, in fact, permitted towns to exempt themselves from the closed-season regulation; so in 1842 the village of Tisbury, on Martha's Vineyard, voted to allow a Heath Hen hunt in early December, provided no dogs were used. (This was the same village that, years before, had voted to protect the birds.) When the fine had been raised to $20, in 1850, Tisbury again exempted itself and allowed a November hunt, without dogs—apparently an attempt to ensure good eats for Thanksgiving. In 1855, however, the state lifted all protections for the Heath Hen—only to reinstate them a few years later. Regulations in other states were equally pointless. The Heath Hen was "ruthlessly persecuted . . . by thoughtless hunters" who defied regulations and faced no serious threat of punishment, wrote ornithologist Alfred Gross.

In 1851, hunter and writer E. J. Lewis bemoaned that New Jersey gunners shot out whole flocks of Heath Hens, "these scarce and beautiful birds [being] butchered long before the time sanctioned by the strong or rather the weak arm of the law." He continued: "Thus it is that the destructive hand of the would-be respectable poacher, as well as the greedy gun of the pot-hunter, hastens to seal the fate of the doomed prairie hen in these eastern regions."

Despite such forecasts and explicit pleas for better laws and enforcement, state governments tragically never fulfilled their own protective regulations, and, at that time, the federal government had no role in wildlife oversight.

Poaching of Heath Hens had, therefore, a long history, which continued well into the twentieth century. In his report on the Martha's Vineyard Heath Hens for 1899, Deputy Game Warden John E. Howland wrote, "I am convinced that more or less [i.e., some] of these birds are shot each year, but as yet I have not been able to get the proper evidence to convict any one." According to Martha's Vineyard resident Charles Brown, writing in 1932, "A few persons shipped bootleg heath hens to Boston, receiving for them $5 per pair," through much of the nineteenth century, in defiance of the law. "This was considered a great price," he noted. Massachusetts's own fish and game report for 1912–14 noted that "the price secured was sufficiently attractive to encourage violations."

Poaching, at least on Martha's Vineyard, may have had more than financial motivations behind it. In the Massachusetts state fish and game report for 1907, we find this curious statement: "It has been even stated that sentiments well-nigh voodoolike in tendency were current on the island, *e.g.*, that a boy must eat heath hen before reaching a certain age."

Collectors also took their toll on the Heath Hen. Well-known collector C. E. Hoyle offered 80 Heath Hen specimens for sale in 1904 at the bargain price of $700 for the lot—and this was but a minuscule portion of the Heath Hen specimens he had procured for individuals and museums. Working in the 1890s, Hoyle shot most of the world's museum specimens of the Heath Hen, doubtless having a significant impact on the already diminishing population. (I once saw several of Hoyle's Heath Hens; one of the specimens, a female, had a chest feather that stuck out so that each time I breathed the feather fluttered.)

In 1905, Vineyard resident and Deputy Game Warden Howland wrote that, about 1895, "There was a craze to get specimens of these birds by collectors and museums, it being understood they would soon become extinct . . . If a man hunted a week and got one specimen he possibly got $20 for it, so he was working a good thing, and the fine was only $20 if caught." Poachers could almost always break even, at worst, or make a considerable profit, at best.

While shooting played a critical role in the decline of the Heath Hen, the bird's decrease also was influenced by far more subtle relationships among the bird, people, settlement and fire.

In 1830, 60,000 people lived in Boston, while New York teemed with nearly 200,000. About 80,000 people called Philadelphia home and the same number held for Baltimore. Some 19,000 people lived in Washington, D.C. Workers laid lines for the first passenger rail service in 1830, while others dug canals throughout New England. Improved transportation provided market hunters greater access both to the hinterlands and to the burgeoning cities. Farmer-settlers continued to push back the limits of the frontier. As an 1867 book devoted to the proper selection of game at public markets said:

> No doubt, in the course of future seasons, we shall have our markets supplied with many of the most choice and rare species of game found inhabiting the distant climates and regions, and placed before our citizens as articles of food. At the same time, the advance of agriculture will be the extermination of these animals from the face of our continent.

The book goes on to provide advice for choosing the best of such animals as Heath Hens, bison and badger.

Curiously, though, the clearing of habitat for towns and farms probably helped the Heath Hen for a brief time. The clearings provided new grounds for leks, as well as places for additional foraging. Heath Hens also learned quickly to adapt to new food sources—crops. "[T]hey perpetrated quite as much mischief upon the tender buds and fruits of the orchards, as well as the grain in the fields," wrote naturalist and sportsman E. J. Lewis, "and were often so destructive to the crops that it was absolutely necessary for the farmers to employ their young negroes to drive them away by shooting off guns and springing loud rattles all around the plantations from morning till night."

Over the years, however, old clearings and ancestral leks began to grow over as settlers pursued a fanatical course of fire suppression. Without fire, the Heath Hens began to lose the brushy habitat and meadows they needed, while at the same time they were being massacred near farms and towns. Fire suppression also allowed tremendous loads of combustible timber to accumulate, making fires more destructive to animals and to humans whenever they did occur.

Altering habitat, in conjunction with relentless overhunting, doomed the Heath Hen—at least on the mainland United States. As early as 1792,

Heath Hens had disappeared from New Hampshire. Humans had extirpated them from the Connecticut Valley shortly after the War of 1812. By that time, Heath Hens had vanished from mainland Massachusetts. On Long Island—a favored habitat—the bird was gone by the mid-1840s. There is no evidence the race lasted on the American mainland much beyond 1870, when Heath Hens had been exterminated from Pennsylvania and New Jersey. From then on, the Heath Hen lived only on Martha's Vineyard, where a flock had flown of its own accord at some time in the distant past.

Local game warden John Howland knew that Martha's Vineyard was the last holdout for the Heath Hen. That made matters all the more frustrating. The closed season instituted in 1890—with only 200 Heath Hens left—had had little effect. In 1896, fewer than 100 birds remained. According to state reports, Howland tried but never succeeded in gathering evidence to convict suspected poachers. Eventually he turned from foiled law enforcer to persuasive lobbyist.

On July 21, 1905, Howland wrote to the Massachusetts Fisheries and Game Commission, urging the members to realize the likelihood that the Heath Hen would go extinct if the state didn't take drastic action and soon. Suddenly, the State of Massachusetts got serious and began to enforce a new, five-year closed season with the threat of $100 fines. Almost as if to underscore the dangers, a fire in 1906 raged on Martha's Vineyard, and the Heath Hen population dropped to 80 birds.

Two years after Howland's letter, in 1907, the Massachusetts Legislature set aside a refuge for Heath Hens on Martha's Vineyard, having acted on a bill introduced by legislator Ulysses Mayhew. The chair of the state Fisheries and Game Commission, the National Audubon Society, and even the U.S. Department of Agriculture had all weighed in, urging an increase in the poaching fine, retention of the closed season and some kind of protection and breeding program, all of which happened. Now the

Heath Hen would have a protected home, a reservation of its own encompassing 1,600 acres within the island's interior.

In 1908, the state bought another 600 acres, complete with a barn and house serving as refuge headquarters. The state then rented another 1,000 acres, and additional money flowed in from private contributions. In a flurry of ethically enlightened behavior, Massachusetts sought to undo decades of malignant neglect and destruction of the Heath Hen.

The legislature outlined key issues to be confronted in the effort to save the bird: the extent and types of plantings on the refuge; protections against poaching, disease and fire; and a breeding program to increase the Heath Hen's numbers. In 1907, the year of the refuge's opening, someone located nine Heath Hen eggs and gave them to a bantam hen to incubate. But on June 20, as a Heath Hen chick began to hatch out, the bantam hen struck repeatedly at the egg until the chick died. None of the other eggs hatched. The same experiment occurred with Ring-necked Pheasant eggs and, in that case, the barnyard hen treated the chicks as her own.

The Heath Hen population by 1908 had fallen to between just 45 and 60 birds on the entire island, despite the fact that 10 broods had been hatched the year before. Then the Heath Hens began to rebound, with 200 birds in 1909 and 300 birds in 1910. The next year, though, the population plummeted to 150. The Heath Hens recovered again in 1912, doubling back to 300 birds. The factors behind these tremendous fluctuations were not clear, because no one had organized a scientific field study of the Heath Hen—the only glaring omission in Massachusetts's plan of action.

In 1913, the state appointed William Day as refuge warden, the caretaker for the some 400 Heath Hens now living on the reserve and elsewhere on the island. Day cut down shrubs and trees to create firebreaks, guarded against poachers, planted winter crops to supplement the Heath Hen's diet and shot creatures suspected of eating Heath Hens: 15 Northern Harriers (then called Marsh Hawks), 3 Sharp-shinned Hawks, 5 Red-

Heath Hens feeding in a meadow on Martha's Vineyard.

tailed Hawks, 23 feral cats (the summer people often abandoned hundreds of mousers upon leaving the island), and 18 rats. Day also planted trees, 7,000 pines in one year alone, as part of an effort to develop a timber economy on the Vineyard—an effort that could not have helped the Heath Hen, which needed meadow and scrub oaks, not pines.

When, in April 1916, the state ornithologist for Massachusetts, the distinguished E. H. Forbush, visited the island he counted 800 Heath Hens on the refuge. No one disputed Day's contention that the birds now lived not only on the reservation but virtually everywhere on the island, numbering about 2,000. On any given afternoon, William Day could roust 300 Heath Hens at once. Finally, it seemed, the effort to save the Heath Hen had succeeded. Strange cackles, moans and booms filled the salty dawn and dusk air. Birds leapt and blustered. Cocks shrieked at others mating. Eggs formed inside the skulking hens. Heath Hen music—notes of Caliban and Ariel—had never seemed so lively and full of promise.

At 6:30, the morning of May 12, 1916, winds blew strongly out of the northwest, nearly gale force, reaching 32 miles per hour throughout the day. That morning someone walking or riding or driving the road that slants northeast from West Tisbury to Vineyard Haven threw away a match or a lit cigarette. The ember fell on a dry leaf, shrinking into a black-and-red line of fire and smoke. Then another leaf caught, then another, then pine needles. The fire surged. It roared through dry plains and woods, covering in four hours the eight miles between the interior town of West Tisbury to the beach community of Edgartown. Cinders blew across the water to neighboring islands. The fire threatened homes, burned stacked cords of wood and destroyed barns, sheds and a gun club.

According to the *Vineyard Gazette*, the "fire wardens and their squads fought the blaze with backfires, or shovelled and plowed great furrows to be used as fire stops. But a shift in the wind or flying fragments of burning woods almost nullified their efforts. The fire stops made by the state along the crossroads proved entirely inadequate . . ." Frances Hamerstrom recalled how, as she played tennis with other well-to-do youngsters at West Chop, a panic-stricken fisherman pulled up to the courts, yelling, " 'Fire! The Great Plain's afire!' He pointed at the empty back of the truck. 'We need help.' " Stunned, no one moved, and the fisherman drove away. "I was ashamed," Hamerstrom wrote. Even as the fire caught in nearby trees, she and the others went back to playing tennis. (Years later, Hamerstrom and her husband Frederick would study with Aldo Leopold and devote their lives to researching and saving the prairie-chickens of Wisconsin.)

Not till darkness did the fire relinquish. Dawn revealed the toll: nearly 13,000 acres burned—about 20 percent of the island. Most of the shrubby scrub-oak barrens the Heath Hens needed for successful nesting had gone up in smoke. Of course, fire created the scrub-oak barrens by vanquishing less fire-tolerant plants. But fires during nesting season—especially where much combustible material had built up for years—that, the Vine-yard Heath Hens did not need.

The buildings on the refuge had been saved. The fate of the birds, however, was more complicated than either outright survival or utter de-struction. The fire destroyed eggs, cover and food—and, some claim, many setting hens themselves. Ornithologists E. H. Forbush and Alfred Gross, and refuge manager William Day, all wrote that hens *stayed on their nests* during the fire, literally being cooked alive.

Certainly hens had been observed to stay close to their well-hidden nests—so much so that humans could even stroke a hen's back. Some-times a person collecting eggs had to *push* a hen off her nest, while she pecked violently at the offending hands. Did fire transform the nests into the hens' graves?

The historical sources argue just that, but offer no explanation for such bizarre behavior. Prairie-chicken biologist John Toepfer believes that two factors account for the impression that hens would have allowed themselves to be roasted alive. First, the fire exposed many Heath Hens already dead for other reasons, such as disease or predation. Second, hens who fled before the onrushing flames probably were caught in swirling winds from which they could not escape. The winds flung the hens back into the blaze.

After the fire, attempts at new nesting did not readily succeed, given the lack of plant cover. Day wrote in his report to the state that "the birds dropped their eggs promiscuously in places where they were soon de-stroyed by crows."

Then, during the harsh winter that followed, there arrived a great nemesis: flocks of sharp-eyed, sharp-taloned Northern Goshawks, one of the most aggressive of raptors. These large, powerful accipiters usually stay in their forested northerly breeding range during the winter months. Every decade or so, however, the populations of their prey—Ruffed Grouse and snowshoe hares—plunge. Then the hungry hawks sweep south, their red eyes spying the most nuanced movement in grasses and trees and snow. Fast, relentless chasers, the blue-gray goshawks set upon the Heath Hens and crammed their hawk stomachs full. The Heath Hens sought refuge in the woods that remained, but to little avail against a predator who, Audubon said, "passes like a meteor through the underwood."

In spring 1917, the plains did not drone with booming, excited males. That year William Day quit his post, perhaps because of the decline of the Heath Hen, perhaps because the firebreaks for which he had been responsible had been criticized following the May 12 disaster.

Of a population that one year before numbered 2,000 birds, there remained only 126 Heath Hens.

Just a few weeks before the May gale, at an April 21, 1916, meeting, state officials and leading ornithologists, including Forbush and T. Gilbert Pearson, had decided to use some Heath Hens from Martha's Vineyard to attempt to reestablish populations back on the U.S. mainland.

Certainly the disasters that followed the April meeting emphasized the special dangers isolated island creatures face. After the fire, Day had urged the relocation of a few pairs of Heath Hens to avoid population crashes brought about by circumstances beyond human control.

So, despite the population plunge, Massachusetts officials went ahead with their plans and captured some Heath Hens. The state dispersed the birds to New York conservation authorities on Long Island and to a game-bird breeder in Wenham, Massachusetts. The hope was that the captive birds would soon begin to breed.

Ornithology's shoot 'em-and-tag 'em days weren't over yet, but this experiment likely marked the first high-stakes meeting of ornithology and poker: a conservation gamble. The experiment failed. Nature called its bluff. None of the birds bred successfully. The captured Heath Hens—16 males and 10 females—died in captivity, representing a substantial portion of the world's remaining Heath Hen population.

In April 1918, a new character emerged in the drama of the Heath Hen, a Vineyard man named Allan Keniston, who took over as superintendent from temporary manager James Peck on April 8. Peck had spent the winter patrolling against poachers, shooting Heath Hen predators and putting out food for the birds, which had flocked onto the refuge on account of a sparse crop of acorns. In addition to the hard work of planting buckwheat, corn and clover for the Heath Hens, tending to firebreaks and keeping both poachers and predators at bay—not to mention making repairs to the state's cold, poorly built two-story residence—Keniston was charged with hosting visitors during his first spring on the job.

One was Norman McClintock, a public lecturer from Pittsburgh who arrived to film the Heath Hen courtship displays. McClintock realized that the sound of moving film would disturb the Heath Hens, so he first placed a loud clock on the field to acclimate the birds to the ticking.

In April arrived E. H. Forbush, who had been conducting annual spring counts of the birds since 1916. Forbush must have been pleased to hear tooting males on the refuge, males who had been virtually silent the spring before. And though he saw only 21 Heath Hens (most of them meandering about in the refuge cornfield) and though he estimated the population at only 150–155, Forbush expressed a cautious optimism:

> With only a small number of birds on the island, with a reservation where they have been concentrated and fed when food was scarce, you have a large food supply in proportion to the number of birds, and this alone, in connection with proper protection from their enemies, ought to insure a considerable increase, provided fires can be kept down for a few years.

Forbush also suggested setting out masses of earthworms during the tooting season; he believed such a diet would bolster the birds' fertility.

That spring, a Dr. A. K. Fisher from the federal Biological Survey also visited the island, accompanied by Keniston and Forbush, in order to survey the system of firebreaks and to offer advice on predator (or "vermin") control. Fisher told Keniston that the firebreaks needed further clearing out of brush and suggested that 18-inch-high mesh fence be installed to prevent embers from blowing free. "Collecting leaves and trash" on the backside of the fencing would help with lighting backfires, if those proved necessary. He suggested that Keniston get help in enforcing game laws and that "windbreaks . . . made of cornstalks" and "strong frames" occupy the boundaries of fields. Fisher also offered advice on trapping cats and hawks, suggesting that Keniston use a tethered domestic pigeon as a lure to bring hawks within netting or shooting range, though I found no evidence that Keniston adopted this tactic. But he did deploy beagle hounds to sniff out feral felines, and, for the next several years, Keniston shot many hawks and cats, trapped them, too—and poisoned rats—all in a diligent effort to check the visible enemies of the Heath Hen. In 1919, Keniston killed 35 hawks, 19 cats and 258 rats (who were suspected of preying on eggs and chicks).

Keniston also visited with farmers and gardeners who claimed Heath Hens ate their crops. He thought their reports were usually overstated and the depredations infrequent, but these rumors of crop damage unquestionably hurt the Heath Hen's already poor reputation among many islanders.

It didn't help that the Heath Hen stood in the way of other game birds. Hunters on the island wanted to introduce the exotic Asian Ring-necked Pheasant, but the state refused, emphasizing instead its desire to protect its rare native chicken. There can be little doubt that poachers not only killed Heath Hens for food, they killed them to make room for other game. (A native Vineyarder familiar with the Heath Hen's history told me confidentially that he thought poaching had remained quite commonplace and that "reportedly there are a few families with skins hidden in their attics" still today.) Keniston's official reports never mention arrests, but one wonders how many tense confrontations he had with suspected poachers and how such work affected the refuge keeper, his wife and son.

Plowing, patrolling, shooting—Keniston did it all, and expertly. Yet 1919 showed but a modest increase in the number of Heath Hens: only about 165 birds. These at least stayed dispersed over a wider area of the refuge and island, reducing chances for a fire decimating them all at once. Keniston saw more broods in 1919 than he had seen the year before, raising expectations for a continuing population increase.

In 1920, he claimed, "Certain it is that the birds are steadily regaining their lost ground." Despite much rain that spring, there were many broods in the summer. The bad weather prevented a proper census, but Keniston could watch 45 to 60 birds feeding together in the spring, a high number for that time of year. He hazarded a guess—that the population had rebounded to "at least" 600 birds. Curiously, no one carried out attempts at capturing and captive-breeding this time, even with the marked upswing in numbers.

The next year Edward Forbush returned for his spring census, actually counting 314 birds—an indication of a higher population across the entire island. (The birds Forbush counted just slightly outnumbered the visitors who came to see the Heath Hen—some 300 who wanted to watch a bird increasingly popular with off-islanders and summer people.) With the apparent increase and dispersal of birds across a wider area, the Heath Hens seemed safer from fire and local flooding—though not from poachers. Furthermore, with the birds scattered off the refuge, an exact and total population count was impossible to come by.

Then a new enemy struck. Three hard frosts in May and one in June. "Severe enough to crack duck eggs," Keniston noted, though he hoped that "the frosty breeding season was offset to some extent by the absence of fires and scarcity of hawks, which enabled a greater proportion of the hatch than usual to reach maturity."

The count for 1922 revealed only 117 birds actually seen and an overly optimistic estimate of—perhaps—500 birds on Martha's Vineyard. The first hint of a downturn in numbers came on January 1, when Keniston looked west of his house, toward the cornfield, and counted only 43 birds, markedly fewer than he'd ever seen at that time in previous seasons. Perhaps the frosts had done more damage than suspected. The Heath Hens "have not increased," Keniston admitted, "to the extent that was to be expected with all the protection and feeding . . ."

The hardworking refuge manager sometimes stared out the window or stood by a tree at the field's edge, straining to see something tangible—a hawk perhaps—something to which blame could be assigned. Why were the numbers again fading? Keniston could not spy the answer. He saw just 43 Heath Hens in a cornfield.

Capawack

AS THE TRAIN LURCHED FORWARD ON ITS TRACKS, LEAVING BOSTON ON APRIL 9, 1923, Alfred Gross knew he had just commenced upon a compelling scientific enterprise. The ornithologist had been commissioned to conduct a long-term field study of the Heath Hen. No one had written a definitive natural history of the bird when it was common. Now—with the world's population of Heath Hens reduced to living on the interior scrub-oak plains of a single island—Alfred Gross would travel there and learn what he could of the Heath Hen's life, demise and possible recovery.

On his way to the ferry at New Bedford, Alfred Gross noticed rivers full of snowmelt and welcomed this retreat of another New England winter. He was temporarily leaving behind his teaching duties at Bowdoin College to spend warming days in a place wholly new to him: Martha's Vineyard, the island that Champlain had dubbed "Vineyard La Soupconneuse" or "The Doubtful Island." The island that, asserted one historian, Native Americans had called Capawack—"the refuge place," "the haven."

As Gross wiped clean his glasses or smoked his pipe, he relished facts from his past that made this journey especially meaningful. Born in 1883 in Mackville, Illinois, Gross was the youngest of seven children. His father, a grocer and farmer, once staked off a bird's nest to protect it from accidental destruction by his farmhands. In high school, Gross collected bird and mammal specimens. He knew Greater Prairie-Chickens, which then occurred in fair numbers in Illinois. Years later, Gross would write an article titled, "Save the Prairie Chicken for Illinois."

A rambunctious boy, Gross hated the muscle-numbing routine of farmwork, even as he savored the access to the outdoors such work provided. He left the farm, entered the University of Illinois and birded—frequently. Sooner than he could have expected, Gross found himself succeeding ornithologist Robert Ridgway as the university's taxidermist. He was then hired to conduct the Illinois Statistical Ornithological Survey—a standardized bird census—and so, with his assistant, Gross *walked* more than 3,000 miles over the entire state in 1906 and 1907, collecting some 200 birds. He went on to earn his A.B. from the University of Illinois and his doctorate from Harvard. While studying for the doctorate, the young scientist met Edna Gross, who was no relation until the two married; the couple honeymooned—on a canoe trip during which Gross nearly drowned.

His Heath Hen research came at a time when the thought of managing a diminished animal back to some semblance of vitality had begun to contest with the morbid practice of collecting all remaining specimens to stuff behind glass. The congenial, rumpled professor (and part-time milk inspector for Brunswick, Maine) would need all of his teacherly verbal skills and his dedication to work in order to negotiate the difficulties that his study would spark. He'd also need his humor and exuberance, which included serving rubber doughnuts to late-arriving students and sudden roadside wrestling matches with friends.

Upon arriving at the island's interior, however, Gross felt little cheer. The center of Martha's Vineyard was an expanse of low-growing trees, shrubs and sandy soil that Gross considered "little more than an uninteresting waste . . . The plains were a study in grays and browns." What seemed to be a barely remarkable landscape sheltered the few remaining Heath Hens on this planet.

Created by channels of glacial meltwater, this subtle, undulant territory is also called an outwash plain. It is a subdued land that might fool the eyes but not the feet. Seeming flat, the plain actually is rife with ravines or bottoms. The meltwater channels are called frost bottoms, because their shaded depths stay far cooler than land just 12 feet higher. Frost can occur in the bottoms in any month; even in July, the temperature in one of these gullies can range over time from freezing to 126 degrees. Far from sterile, the wrinkled scrub-oak barrens on the sandy outwash plain can be biologically fecund.

Perhaps not surprisingly, as Gross began to traverse the land, as he walked paths and stooped and turned among the wild branches of scrub oaks, his sense of the barrens began to change. He heard Fox Sparrows and Eastern Phoebes. Two days after arriving on the island, on April 11, 1923, Gross walked alone in the frosty dark. He was about to encounter Heath Hens on their Vineyard booming grounds for the very first time.

April 11, 1923, 3:30 A.M. It is very dark and only the faint light of the stars illuminates my way to the blind located in the western meadow about one fourth of a mile from the reservation house. There is a heavy frost and the grass crunches under my feet as I walk along. All is quiet as far as the voices of birds are concerned.

3:40 A.M. I am now seated in the blind waiting for the birds to appear. A fog has been coming in and at present is dense enough to hide the stars.

3:55 A.M. First dim light of dawn. I hear the first clear whistled notes of a Bobwhite . . .

4:05 A.M. The first Robin begins chirping . . .

4:21 A.M. The first toot of the Heath Hen is heard in the distance near the western margin of the meadow. It is not yet light enough to see and the fog obscures the view.

4:24 A.M. The toot is followed by hen-like calls resembling cac-cac-cac . . .

4:50 A.M. It is growing lighter. One of the birds has flown to the roof of my blind where I have placed corn to attract them. The blind apparently was first chosen by the birds as a vantage point from which to view the field and as an admirable place to conduct their stamping and tooting performances.

4:53 A.M. The bird on the roof is now eating corn. The two other birds which are near the scrub oaks on the south are busy feeding but are gradually making their way toward the blind. Their feeding, however, is frequently interrupted by their tooting or so-called booming whhoo-ooo'dle-dooooooh! *followed by the hen-like,* cac-cac-cac *or* oc, oc, oc, goc, goc, goc, *occasionally ending with a queer call which might be represented by* auk, a-e-e-e-e-ek *and sometimes it is more nearly* e-e-e-ek-ek-ek-ek. *Frequently one of them leaps into the air to a height of three or four feet and in so doing utters a loud piercing* urrrrrrrb *followed by a curious laughter-like sound. In this wild demonstration the bird completely reverses its position so that it faces in the opposite direction from its initial position when it alights on the ground. This leaping and screaming seems to be augmented by similar performances of the members of the group somewhere in the western part of the meadow.*

4:58 A.M. *The bird on top of the blind has just stamped its feet and tooted and while so doing I could clearly see it through a space between the roof boards.*

5:00 A.M. *The sun which is well above the horizon has just appeared through the fog as a great red ball. The Quail have ceased calling, the chorus of Robins has diminished but the Song and Vesper Sparrows are singing as vigorously as ever . . .*

5:02 A.M. *One male from each group runs rapidly toward another male in a defiant war-like attitude. When near together they hesitate, lower and waver their heads, leap at each other and strike their wings vigorously as they leap. Apparently no harm is done and they settle back into a comfortable position facing each other.*

5:03 A.M. *One of the birds has arisen, has run away a short distance and is now going through his tooting performance without any interference from his antagonist.*

5:07 A.M. *Something has frightened the birds, though I cannot determine what it is. The bird on top of the blind and all except one of the birds on the ground have flown from the field to the scrub oaks.*

5:16 A.M. *Two birds have returned from the oaks . . .*

5:48 A.M. A Heath Hen has flown to the top of the blind . . .

5:55 A.M. The bird on top of the blind is pecking at the corn and makes a great deal of noise as the ears are rolled about. At irregular intervals whenever the occasion seems to demand, the bird goes through his tooting performance; at other times he merely rests quietly near the edge of the roof apparently interested in his fellows or the general landscape . . .

6:30 A.M. The sun is brighter and the fog has been completely dispelled, but a chill wind has started up. One bird still remains on top of the blind and I am getting some very fine views of him through a reflecting mirror offering unexcelled opportunities to study its vocal mechanism. The stamping which precedes the tooting and which seems like a very minor part of the courtship performance when it takes place on the ground, is one of the chief features when the resounding pine boards directly over my head intensify the sounds . . .

7:00 A.M. Now that the excitement of watching the Heath Hen[s] is over I realize how stiff and cold I am even though I am dressed in the heaviest of winter clothing.

7:30 A.M. I remained in the blind until 7:30 to make sure the birds would not return. After being away from the field for this length of time experience has shown that the birds are not likely to return again until the afternoon, especially when there is such a strong sharp wind. I left the blind feeling cold but well-rewarded by the rich experiences of the morning.

Soon enough, Alfred Gross remarked, this land presented "a lure and attractiveness about it that I could not at first appreciate." *Lure and attractiveness*—that is what a landscape, an ecosystem, a watershed can offer, if one stays put long enough with diligence, senses ready for anything. Even a wounded place, bereft of a creature or losing one, can be a salve. Capawack. Refuge. A haven. From the blind on the hill, in June, Alfred Gross could watch a stunning meadow of wind-waved daisies. He admired white campion, frostweed and tall crowfoot. He heard Eastern

Alfred Gross holding a Heath Hen on May 16, 1923.

Bluebirds and Brown Thrashers. He heard Heath Hens and pinkletinks—tree frogs—and, for him, the island had come to life.

Gross returned from his scrub-oak-and-meadow forays to the refuge manager's house, the home of Allan Keniston. This unadorned, L-shaped two-story became Gross's home-away-from-home. A ramshackle, echoing affair, the Keniston house still stands today, replete with office desks, telephones, posters, an old clock, signs leaning against the walls and one antiquated yet stylish black manual typewriter. The renovated cow barn now functions as a kitchen, and upstairs are two bedrooms. The house serves as the headquarters for the Manuel F. Correllus State Forest (where President Bill Clinton once jogged while on vacation). Today ragged stands of white pine, pitch pine and spruce dominate over a few patches of open field and waist-and-chest-high scrub oak. Looking from the old Keniston house, which I visited in 1998, I found it difficult to visualize the area as primarily an open meadow and scrub-

The Keniston house—and Heath Hen Reservation headquarters—as it looks today. The building serves as offices for the Manuel F. Correllus State Forest.

oak expanse that, in spring, reverberated with the booming calls of Heath Hens.

While Gross continued his observations, Keniston kept busy with spring crop plantings and road repairs. The work could not distract them from the dismal numbers, though—numbers far below Keniston's inflated estimate in January of nearly 150 birds. On May 3—a time when the scrub oaks had begun to be festooned with "hundreds of Van Dyke–red catkins"—Gross's census count revealed only 46 Heath Hens. This was far better than the April count of 28 birds. Yet the population teetered.

So bleak appeared the outlook that the Massachusetts Department of Conservation's Division of Fisheries and Game mailed a survey to some 100 individuals and organizations, including Martha's Vineyard politicians, nationally respected ornithologists and private donors to the state Heath Hen refuge. With a land lease for the refuge up for renewal, Massachusetts officials wanted to know: Should they cease their efforts to save the Heath Hen?

The language of the 1923 annual report strongly suggested that the state took seriously the option of ending help for the Heath Hen altogether. In part, the report noted that the effort "apparently [was] a losing cause" and funds expended for Heath Hens—$3,000 to $5,000 annually—might be better put to "stock[ing] the island with other varieties of game, from which perhaps a greater portion of the tax-paying public would derive benefit and pleasure." Fortunately, with 49 of 65 replies answering a resounding "No!"—don't give up—the state renewed its lease and the effort to save the Heath Hen continued.

But the survey had not fully sampled opinions on the island concerning this bird, which some called the "heth 'en," as if it rhymed with "brethren." Gross found many islanders who felt, at best, nonchalant about the bird and who were sometimes, openly hostile. The *Boston Sunday Post* for April 29, 1923, included a comment from a Vineyarder who wondered if it was appropriate to attempt to "propagate a bird that fails either to feed or clothe us, or enjoy our society; unless, like the sacred ibis, it is to be maintained for the purpose of worship." As early as 1906, the *Vineyard Gazette* ran comments from islanders who thought, "Why not let [Heath Hens] go now since they are almost extinct. Then we can class them with the dodo while we are alive, perhaps?" In spite of such disdain for the bird, Gross called on "leading citizens" to become involved in the effort to save the Heath Hen.

Following the survey, Division of Fisheries and Game Director William Adams seemed to throw himself fully behind the effort to protect and restore this bird, but in later years the fact that the Division of Fisheries and Game had even considered closing the refuge would be used against him and the state. It didn't help the state's image that officials claimed it could not afford to fund Gross's study; monies for his fieldwork came from private donors, including John C. Phillips, who had tried to breed Heath Hens following the 1916 disaster. For the next five years, Gross would be funded by the Federation of New England Bird Clubs, Inc. Blame for the state's failure to fund Gross's work would accrue to Adams, though the professor specifically noted that Adams had fought a losing battle for funding with the Governor's office. While not providing direct support for Gross, the Division of Fisheries and Game admitted that it would take management cues from the results of Gross's findings. The ornithologist would become the de facto—and high-profile—director of state Heath Hen policy.

After the disappointing spring 1923 census—and a failure to locate any females at all that season—Gross and Keniston were pleased by the later reproductive success of the flock. Weeks of nesting in good weather produced 11 chicks in three separate broods. This was measurable progress, albeit only a little. That year Gross also trapped five males in order to study them in captivity and to lower the ratio of males to females, the latter being severely outnumbered. All five died, but the bodies were saved for autopsies.

In July, Gross and Keniston nearly killed a hen and her chicks when, driving during a rainstorm, the two men came upon the birds brooding in the middle of a dirt road—an odd place to be during a downpour. "One [chick] flew against the radiator," Gross wrote, "then disappeared from view." The birds all flew away. Because the deep ruts of dirt roads were favored places for Heath Hens to take dust baths, any number of the birds may have been killed by less-alert drivers.

The year 1924 was the first in which Alfred Gross began to shape state policy toward the Heath Hen. Massachusetts agreed with the professor that Keniston should discontinue extensive crop plantings that—in the moderate winter of 1923–1924—did not attract Heath Hens, which had plenty of wild foods available for foraging. The crops, instead, had attracted crows and rats, some of the very creatures Keniston had been killing because of their perceived attacks on chicks and eggs. Winter feeding would now involve buying and setting out stocks of seeds that the Heath Hens might eat, and such feeding would be carried out wherever Heath Hens lived, even off the reservation.

Relieved of the most burdensome farming duties, Superintendent Keniston now helped Gross with Heath Hen observations, monitoring not only the reservation flock but any birds living on private land. Keniston also continued shooting cats, hawks and rats, and patrolled the land to prevent poaching. In April 1924, Gross counted 54 Heath Hens—a slight increase from the previous May—and that summer Keniston reported at least 4 broods of chicks, including one large group of 10, despite wet and cool weather poorly suited for raising young. The broods raised hopes that soon there would be a significant upturn in the population.

Gross trapped four cocks, again trying to lower the number of males to females, there being never more than three females that he could discover at any given time. Careful not to remove more than five males in a year—to avoid traumatizing the flocks—Gross knew that the operation entailed risk. Yet he was concerned that too many males were vying for too few females. "The males may molest the females to an excessive degree, tending to produce sterility . . . it is conceivable that a dozen bachelor males would play havoc with a female tending a brood of downy young." The Heath Hens needed more females so that more eggs could be produced, and the females that remained led haggard and vulnerable lives.

Another kind of havoc appeared in 1924. That January, a sportsman and activist named Lloyd Taylor wrote to Alfred Gross and criticized the state for its "inefficient care of these birds." Taylor, like others, continued to harp on the matter of the 1916 fire. He said that "efficient, careful men should be in charge of what remains," a not-so-veiled attack on everyone from Fisheries and Game Director William Adams to the man on the ground, Allan Keniston. Taylor remained critical in a November

28, 1924, letter, implying that the state's "vermin control and fire guards" remained inadequate; further, he opined that the introduction of female western Greater Prairie-Chickens be considered.

Of Taylor's proposal to import western-race prairie-chickens, Gross felt dismissive. He rejected it out of hand, noting that many people preferred the Heath Hen to die out, if it must, as a "pure" race of birds and, anyway, he reminded Taylor, reintroduction of Heath Hens to the mainland had failed miserably. There seemed little chance that importation of Greater Prairie-Chickens to the Vineyard would fare any better. After all, a release some years before on the island of a handful of western-race Greater Prairie-Chickens apparently also had failed, with those birds simply let go into unfamiliar territory where they soon died.

Critics of the Heath Hen policies had found, though, two of their most powerful attacks—state "neglect" and a demand for cross-breeding—both of which they would volley again and again in the months and years to come. Less than a year after he had begun his visits to Martha's Vineyard, Alfred Gross began to understand, if he had not done so already: While the fate of the Heath Hen constituted a problem of biology, the management of biological issues became a public concern. And public perception of the Heath Hen's fate was, at least in part, a matter of politics.

When, before dawn on May 12, 1925, Alfred Gross left Allan Keniston's house, it must have been with a sense of deep foreboding. After a moderate winter, the Heath Hens should have been doing well, especially given the number of broods in the previous spring. Yet only 31 birds had shown themselves in early 1925. In April state ornithologist E. H. Forbush had been on the island, observing just two birds, both away from the reservation. The drumming fields stayed silent, a sobering and astonishing absence. Even as Keniston built a new blind for Heath Hen observations, it seemed as though the birds, like meltwater, were trickling away into time's hidden rivulets.

In the dark, Gross walked from the trees that lined the path between the house and barn and headed toward the meadow—the heart of the lek. No birds sang from the shelterbelt of trees on the property's north side; it was too early for song. He continued past the duck pond on dewy grass. Gross walked west, a quarter of a mile from the house, to reach the field that served as the Heath Hens' lek and as a feeding ground. This was at the top of a small hill or rise. (Today, an Eastern Bluebird nesting box marks the location. Local lore has it that the foundation of an old fire tower, farther west yet, was the site of Gross's blind, though it may have been moved from one area to another.)

May 12, 1925, was a lovely day—the air became warm and pleasant—but Alfred Gross found just two Heath Hens on the refuge property. At James Green's farm near West Tisbury, there lived just one Heath Hen, who had probably always frequented the farm in a group separate from those that preferred to stay on the reservation. For a while this lone male constituted the entire "flock" on Green's property.

Three birds seen. All males. Gross estimated that there were just 25 Heath Hens living on Martha's Vineyard.

While humans worried over the grim numbers—the population had virtually been halved in just one year—some unseen females managed, almost miraculously, to raise chicks in three separate broods. Keniston watched a half-dozen chicks gamboling in a field between the refuge fire tower and the house.

Then the rains came. More than two inches of rain fell in 72 hours, drowning the chicks, washing them into road ruts and muddy puddles or else chilling them beyond a saving warmth. No reliable observers ever saw Heath Hen chicks again.

The same month that the chicks vanished, Alfred Gross spoke to about four dozen people gathered in a conference room at the State House in

Boston. It was June 4, 1925. On that day, some cultural threshold was passed. In years prior, news of the Heath Hen crisis would have set off a fierce round of shooting, the last frenzy of museum and private collecting, to get the rare and valuable skins. Not this time. Science had taken a stand against collecting and urged conservation management instead. The conference participants—officials from the Federation of New England Bird Clubs, the State of Massachusetts, Massachusetts Audubon and others—drew up a list of desperate emergency measures to stave off further losses.

A new state warden would be hired to patrol the refuge from October to March—"a complete patrol . . . rigidly enforced," an edict that suggests winter poaching had persisted to some degree. Additionally, James Green, the farmer, would be deputized as a warden, because of the Heath Hen present on his property. A third warden, Edward F. McLeod—to be paid for by the Federation of New England Bird Clubs—would be hired to shoot "vermin" and assist in patrols; that organization immediately began to raise funds to supplement the state budget. Feeding and baiting of the birds would continue, in order to encourage their presence on the protected reservation ground. Finally, the Federation paid Pennsylvania's Chief of Vermin Control, J. J. Slauterback, to offer advice on how best to trap, poison and shoot hawks, rats and feral cats.

All of these steps came at a time when the Heath Hen had become something of a star. Newspaper and magazine stories had sensitized people across the country to the plight of this bird. Local opposition to the Heath Hen appeared, over the last two years, to have moderated, at least publicly. As part of the effort to win over opinion, Allan Keniston delivered lectures to the Y.W.C.A. and talked with some 100 visitors to the Heath Hen reservation.

Sympathy and wardens, traplines and rat poison—all the sentiment and activity continued to galvanize public interest and, perhaps for the first time in American history, there seemed to be a developing consensus that an attempt to save a threatened creature could be a worthwhile, even noble, goal.

One suspects that Alfred Gross understood all this as he walked out of the State House conference room late in the day on June 4, but that he also felt grim. Because the bird seemed, at that moment, beyond saving. In the state report on the Heath Hen Reservation for 1925—in a passage emphasizing the "intensive trapping campaign" and the examinations of the stomachs of hawks and other vermin—there occurs this telling admission: Not one autopsy showed Heath Hens in those stomachs. Of course hawks did eat Heath Hens, but this was evidence that hawk predation mattered less in the demise than previously thought. Further, the report noted a marked rise in the Bobwhite Quail population at just the time the Heath Hen had decreased. This, the report said, "would seem to indicate that the decrease in the heath hen may not be attributable to vermin, of which there has been no increase during this period." A subsequent study concluded that rats typically did not visit habitats frequented by Heath Hens during the birds' nesting season.

The vermin-control campaign nonetheless stayed in place, seeming to Gross more and more feckless. Yet Massachusetts hunters—especially those on the island who had so recently taken an interest in saving the Heath Hen—began to focus great energies on this campaign. It embodied, if nothing else, a tangible effort. One could count how many vermin had been shot. And so McLeod got to work, dispatching 20 rats, 26 crows, 4 owls, 13 hawks and 54 cats. Keniston's tally for 1925 included 119 rats, 12 crows, 30 cats and 23 hawks.

There remains, in the old Keniston house, a scrapbook with pictures of the place and the work—including a photo of a trapped Northern Harrier caught eating a Heath Hen, the shocked hawk looking back over one wing while the Heath Hen is nothing more than a white sprawl of overexposed light, a terrorized glare without substance. I saw a grisly picture of dozens of dead cats hanging from strings—summer mousers killed as vermin. Also in the photo album, a handwritten entry in white ink on black construction paper: "The heath hen as I saw them." The picture is missing.

Gross kept thinking about dead bodies—dead Heath Hens found three years before by Keniston, as well as by Vineyarders Frank Goulart and Harry Athern. The corpses showed no signs of hunting wounds. Now, with the population plummeting, Gross suggested to the state that disease had been at work. While Keniston augmented his traplines on the reservation and elsewhere and McLeod intensified his vermin-shooting campaign, Gross returned to his teaching duties at Bowdoin College, worried that something was invading the bodies of Heath Hens going about their daily feeding and loafing.

McLeod and a hunter-activist named Arthur Clark felt convinced there were more Heath Hens than the professor believed, so they took comfort in the dozen birds that flocked onto James Green's farm in November 1925. A dozen birds here, McLeod and Clark reasoned with confidence, surely meant more elsewhere.

Although Keniston resorted to planting some spring crops in 1926 in an effort to lure Heath Hens to the protected ground of the refuge, none ventured there. Finding the birds was becoming difficult. Gross had included Keniston, McLeod and Clark in his Heath Hen census team of March 1926, but even with the additional eyes, they saw just 15 birds. Gross estimated only 35 on the island.

But in *his* autumn summary of Heath Hen sightings, McLeod—the Federation of New England Bird Clubs warden—seemed to accept uncritically a few secondhand observations, including those of young birds. McLeod does not contextualize the alleged sightings—he does not evaluate their veracity. To some, this might make the reports seem more credible than, in fact, they were. In a letter of December 7, 1926, to Adams, Keniston wrote that he doubted the *one* brood report he had received. He questioned McLeod's rationalization that so few Heath Hens had been observed because rabbit hunters had gotten a late start on their season and

Allan Keniston in 1926.

therefore had not pushed Heath Hens out of the adjacent state forest and onto the refuge proper.

By late 1926, the situation among the key Heath Hen players had become fractured and complicated, with charges and countercharges hurled in a series of letters, filed in the Gross archives at Bowdoin College. The catalyst was McLeod, who had killed a locally rare Ruffed Grouse and then shot two Eastern Screech Owls—birds with utterly no connection to Heath Hen predation. When he got the news, Fisheries and Game Director Adams grew furious, and on December 4, he wrote to Arthur Clark, who supervised McLeod for the Federation's Heath Hen Committee. Adams wrote that McLeod should have "set the example for the gunners on Martha's Vineyard and refrained from shooting any ruffed grouse," and that, while having a permit from the state for scientific collecting, "it would not appear that in killing the [screech owls] he had in mind any contribution to science." McLeod's behavior, if emulated, could "result in a practical withdrawal of protection" to wildlife, Adams said. This censure seemed to do little to alter McLeod's behavior for, later, Keniston mentioned to Gross something he called "the black duck episode"—apparently another shooting by McLeod of birds with no relation to the fate of the Heath Hen.

Clark, who seemed to relish his developing role as a kind of anti-state insurgent, immediately counterattacked. He solicited positive letters about McLeod's work, including one which claimed, amazingly, that McLeod alone had been responsible for better enforcement of hunting regulations on Martha's Vineyard. Clark convened a meeting of the Federation's Heath Hen Committee on December 8 to discuss a response to Adams's letter. Four members of the committee, along with Gross and McLeod himself, attended the conference in Boston State House Room 136. The committee passed resolutions supporting McLeod and deploring "a regrettable lack of the fullest cooperation among the wardens on Martha's Vineyard." (Whether Gross voted on this and other resolutions is not clear.) Both McLeod and Clark felt that another state warden, Karl Eckert, had snitched to Adams about the screech owl shootings, though, in fact, Eckert simply hadn't known what collecting permits McLeod held and needed to verify with the state what McLeod could, and could not, shoot.

Next, Clark got busy writing Gross—two letters on one day, December 20, 1926—in which he mixed tones of friendliness and presumption, nearly lecturing the ornithologist about the need for all to "be careful" in public statements concerning the Heath Hen. And Clark even criticized Gross's manuscript on the bird. The early drafts of *The Heath Hen*—to be published in 1928 as part of the Memoirs of the Boston Society of Natural History—not only discussed ornithology but politics as well. Gross named names, and Clark took exception. He wrote that "ten years from now or in less time almost no one will care who did this and who said that etc. Also it now seems apparent that whatever you say or whoever you mention in particular someone else will be offended." Clark suggested that all such material go "in the appendix which might be printed in small type all the petty dickerings and scrambling for credit, criticisms if you wish, etc., [which] could be melted into the pot and tacked on to the bitter end."

Alfred Gross decided to leave all such material out of his book. Historians have only a scattered record of newspaper articles and archival material, a partial account of the fights, arguments and claims that show us the human side of the Heath Hen story. The movements of the Heath Hen and the hardships of the heart both marked the relationships of men fighting to protect the bird and remind us that conservation becomes always more than a stated goal, always more than science.

A Phoenix on the Vineyard

...And the end and the beginning were always there...
—T. S. Eliot, "Burnt Norton," *The Four Quartets*

ONE COULD SAY THE END BEGAN WITH EXILE, THE EXTERMINATION OF MAIN-
land Heath Hens and the banishment of remnant kin to an island. Or perhaps the end began more emphatically on the island itself, in 1916, when
the facts of fire and goshawks displaced the possibilities of eggs and flocks.
Images swirl like cinders on a breeze—bough-houses, hands hefting bodies beside a market stall, brown feathers, char, oak leaves.

The last beginning: 1927, a year when the Heath Hen population
again dropped, but public agitators such as Arthur Clark persisted in saying the birds could recover, if only the state would listen to his calls for
more vermin shooting.

Alfred Gross knew better. In both his public and private statements the
ornithologist betrayed a fatalism about the Heath Hen race; he felt nothing could save the birds. Privately, he expressed fatigue with Clark, Edward
McLeod and their supporters. Clark continued to issue public statements
about government neglect of the Heath Hen, almost as if the fate of the
bird was secondary to whatever score it was Clark had to settle.

The foremost reason for Gross's bleak assessment was the confirma-
tion that the birds were indeed showing signs of disease. Studies found
Heath Hens with tumor-like growths caused by the blackhead parasite, a
disease of domestic poultry communicable to Heath Hens, and infections

of the gizzard caused by the *Dispharynx* parasite, which also affects Ruffed Grouse.

Aware that a do-nothing approach was politically infeasible, Gross advised the state to remove all domestic poultry from the Reservation, even though Heath Hens had all but disappeared from the public land. Allan Keniston felt angry and betrayed. He wrote to his boss, William Adams, in Boston:

> Dr. Gross speaks as though the birds have only disappeared "from the Reservation." I know that the birds disappeared from practically the whole range on the Island at the same time. He sure cannot lay that to MY poultry. Most of the birds that were found sick or dead were a long way from the Reservation.

It was late February, and the year already was off to a bad start for Gross. He had confirmed his suspicions of disease and now had inadvertently alienated his friend Keniston. Gross immediately wrote, "My dear Allan:— . . . I did not intend to infer [sic] that the evidence of disease we have recently discovered was due to any fault of yours or of the state department. The reservation was mentioned . . . since it is the only place over which we have control . . ."

By the time of the March census, Arthur Clark no longer chaired the Heath Hen Committee for the Federation of New England Bird Clubs— it's not clear if he quit voluntarily—but Gross invited him, along with Keniston and McLeod, to help with the count. Including Clark and McLeod probably was a last-ditch effort to convince the men that there were far fewer birds than they thought existed.

The men counted 13 Heath Hens.

Ill-timed fire. Plunderous goshawks. Unquantifiable, suspected predation by other hawks and cats. Rains. Frosts. Poaching. Suspected sterility among males. Phone lines that tangled up the birds. Too many males. Too few females. Inbreeding. Disease.

Thirteen Heath Hens seen, no more than 30 conjectured.

With Arthur Clark bent on assigning blame to the state for this decline—and refusing to acknowledge that the population was as tenuous as it was—and with the state dedicated to keeping the remaining birds alive in a wild condition for as long as a Heath Hen lived, Alfred Gross applied his tact as best he could to an immediate matter, one that he could still influence: the status of the trigger-happy Edward McLeod. Gross told the Federation of New England Bird Clubs that their warden had done all he could in shooting vermin (this was damning with faint praise). Then Gross explained that the Heath Hen's decline resulted from a complicated tangle of causes over which no one had control, so it would be prudent to let McLeod go (already having spent $7,000 to support his work). The state, the professor concluded, should remain the Heath Hen's sole custodian.

Thirteen Heath Hens fed among sweet fern and blueberry. They idled in James Green's fields as the farmer stirred a pot of soup or wrote down a list of the month's groceries. The birds continued their ancient rhythms. The humans bickered. Apologized. Then bickered some more.

April 12, 1927, Arthur Clark to Alfred Gross: *"Sorry we had a little 'spat' over the telephone, eventually you will come to realize that my interest is not in politics, as you suggest, but in the heath hen and fair play. Apparently, I can't talk about that bird even yet without scrapping, so I'll quit, but I feel very deeply and sincerely about it."*

April 14, 1927, Alfred Gross to Arthur Clark: *"I am extremely happy to receive your letter and to know that a difference of opinion does not sever a friendship which I prize very highly.*

"...Arthur I honestly think there are not more than the 13 birds otherwise I would not have said so. Your interpretation is different from mine but please don't accuse me of playing politics and I certainly with draw my ac-

cusation ... Now that the Federation has withdrawn we are depended [sic] on the state for the protection of the birds and since we are both primarily interested in the preservation of the birds let's both pull together to that end ..."

The Federation of New England Bird Clubs had dropped McLeod, though he now had the backing of the Martha's Vineyard Rod and Gun Club and its own Heath Hen Committee, chaired by the ever-busy Arthur Clark. In May, the state refused to grant McLeod a permit for continued shooting of owls and hawks, sparking yet another press controversy.

Meanwhile, sportsmen blamed state managers for "domesticating" the Heath Hen through feeding plots on the refuge, even though the last flock lived on the Green farm. They grew angry at alleged statements by Gross that islanders did not want the Heath Hens. As well, the ornithologist and others worried that, by shooting hawks, the wardens might be allowing some feeble Heath Hens to live longer than they otherwise might, perhaps weakening the entire population.

In the summer of 1927, in lieu of broods, there were newspaper and magazine articles, including the cover story for the July issue of *Hunting and Fishing* magazine. Touting the need for continued vermin shooting and twice contemptuously characterizing the state's Heath Hen efforts as "spasmodic," the unsigned article asserted that the bird "has faced extinction before *and survived* when the odds against it were far greater than they are today." With high praise for the Martha's Vineyard Rod and Gun Club and hyperbolic concern about islanders being alienated by the state, the editorial placed the population at just under 50. Such an unrealistic assessment further damaged what credibility Clark, McLeod and their cohorts still possessed. Not once did the piece cite the world's authority on the Heath Hen, Alfred Gross. But readers could obtain a color Heath Hen print for a modest donation.

The anti-state faction failed, however, to win new supporters on the Vineyard and soon began to lose former ones. Keniston observed all this with bemused disdain. "Very few people pay any attention to what Clark or McL. put in the paper," he wrote to his friend Gross. In October, Keniston told the professor that "the *calliope* . . . is holding forth to all that will listen, and I understand the Rod and Gun Club at Oak Bluffs have doubled their dues to raise some money. [I]f this is true I see where a great exodus will take place from the club as it is quite widely known what the raise of dues is for (McLeod)." Later that month Adams wrote Gross to say that he knew nothing of the Club's work, if, indeed, there had been any work at all. Apart from shooting and fundraising, it's not clear just what the Club's Heath Hen Committee actually did.

By the end of this troubled year, Adams had earned the hidden ire of another Vineyarder: his own Heath Hen superintendent, Allan Keniston, who never received a promised raise after he had eliminated his fowl and poultry from the refuge. Keniston was having a hard time making ends meet. He wrote bitterly to Gross on December 16, 1927: "When the reclassification took place most state employees got a raise but I was demoted, put in the class of bird refuge keepers so I will never get anymore than I get now. Kind of them wasn't it? I guess they misnamed the Department anyhow. It should have been the Constipation Dept. instead of the Conservation Dept."

Throughout that winter, the farmer James Green kept looking out his windows. He turned his calendar to the New Year, January 1928, and scrutinized the small flock of Heath Hens feeding in his field near the barn. Where nine had gathered in the autumn, now there were seven. From nine to seven to five went the Heath Hens. The flocks that seemed so small just a few years ago—two dozen Heath Hens, say—must have seemed in retrospect like a version of abundance. Green noted how one of the five birds stayed alone, listless, apparently sick. It vanished. To four went the Heath Hens.

When, between April 5 and 9, 1928, another census took place, the counters outnumbered the counted. Local interest in and support for the Heath Hen had never been higher, and mistakes yielded alchemic conversions of one species to another. Transformations of Bobwhite Quail into Heath Hens were never more numerous. Readers delivered photos of alleged Heath Hens to newspaper editors, who politely explained that the birds pictured were Ruffed Grouse. Heath Hens were everywhere! But the Heath Hens were three, all males.

Olin Sewall Pettingill and Thornton Burgess could see the pitiful trinity through the slot in the wooden blind's wall. The birds were feeding in the late afternoon light of April 6, 1928, on James Green's farm. Pettingill, a student of Gross's, and Burgess, a popular children's author and friend of the Bowdoin professor, were in the blind to take still and moving photographs of the birds. A visual record of the remaining Heath Hens had become a priority, but the existence of these historically important films seems largely to have been forgotten.

At the Westborough office of the Massachusetts Fisheries and Wildlife agency are three Heath Hen films, all silent and stored in a cabinet in a non-air-conditioned hallway. One summer morning, biologist Jim Cardoza and photographer Bill Byrne set up a projector and ran the movies for me. It had been years since they had shown them. One film appeared to be a product of the Pettingill-Burgess collaboration—briefly three male Heath Hens appeared. That film also included a shot of a house and outbuildings right by the Edgartown–West Tisbury Road—the old Green farm. In other films, I saw Heath Hens feeding and Keniston setting cat traps. One reel showed the eminent E. H. Forbush so discombobulated by his brief role as a movie star that he looked through the wrong end of his binoculars. I saw scenes of Heath Hens displaying: The film's caption declared, "Showing the white feather—but not defeated." When I reached Olin Pettingill by phone at his Texas residence, hopeful that he could tell me more about the films—or any other stories—he declined, saying, "My mind is empty. I've nothing more to add ..." Pettingill apologized, graciously referring me to the memoir of his ornithological career, which had begun under the tutelage of Alfred Gross.

For spring surveys in 1928, Pettingill had joined Gross, along with

Burgess, Keniston and warden Karl Eckert. (Clark and McLeod weren't along this time.) The team even canvassed islanders in an effort to learn of more Heath Hens. It had become clear to most, however, that reports of Heath Hens in this or that locale simply meant the three remaining birds had flown from one area to another.

Keniston already had begun filing daily reports on his work routine and on the many rumors of sightings. He now regularly questioned people who claimed to have seen Heath Hens; Keniston once refused to pay a man who demanded $10 just for *saying* that he had spotted a Heath Hen. The superintendent could not simply dismiss reported sightings, though, because public goodwill still mattered. He needed to be thorough.

All that year Arthur Clark continued to vent his anger. In the January 27, 1928, *Vineyard Gazette,* for example, he spoke again of the state's refusal to allow Edward McLeod to shoot more hawks and owls. Clark also complained that the low census numbers made fund-raising difficult and still asserted—outrageously—that the Heath Hens could recover to a flock of several hundred. Clark vexed Gross with acerbic statements, even in letters containing warm, collegial words. In a letter of February 20, 1928, Clark invited the ornithologist to stay with him and his wife in Boston, but remained contentious: "Each of us thinks the other is both misled and prejudiced. . . . I shall expect you to give . . . consideration to my 'layman's' views, and that you have not always done . . ."

Following the publication of Clark's sarcastic and hyperbolic essay, "Lo, the Heath Hen," a friend of Gross's suggested "that young gentleman, I think, needs a good dose of morphine."

By the summer of 1928, much of the state forest—technically not part of the reservation—remained closed to rabbit hunters, a restriction meant to prevent disturbance to the Heath Hens. Poultry and fowl no longer lived on the reservation. Keniston put in additional plantings meant to lure back the Heath Hens and he created new firebreaks. Meanwhile, the Martha's Vineyard Rod and Gun Club refused to offer an account of its activities, though Gross had specifically asked for one. On June 21, Keniston saw only a single Heath Hen, as the bird moved close to the forest's fire tower.

Although the state bill for its Heath Hen efforts totaled nearly $64,000 by 1928, the Massachusetts Division of Fisheries and Game continued to affirm its commitment to protect the Heath Hen so long as one existed. Adams remained as chief of the division, the Governor having reappointed him in June, a move that discouraged the sullen duo of Clark and McLeod.

And, in early May, one of the three Heath Hens had died. Edward McLeod found the corpse "44 paces" from James Green's barn, "in amongst some blackberry bushes . . . It might have been killed by flying against telephone wires." McLeod told only Clark about the death. State officials didn't learn of the loss until summer, after McLeod responded to a specific request from Adams for any new information.

Sixty-two-year-old James Green—farmer, caretaker for the Tisbury Pond Gunning Club, former whaler and ex-ferryboat-fireman—watched the two remaining cocks in his field that autumn.

After December 8, 1928, he saw just one.

Allan Keniston bent to look at the tracks in the February snow. Grouse? No. *Heath Hen*. The tracks meandered just west of the fire tower, giving evidence that the lone bird was recently alive. With a broad expanse of scrub oaks in which to hide and feed, the world's last Heath Hen had kept well hidden in the winter months. So well hidden, in fact, that Keniston joked the bird probably was waiting at the post office for letters from Edward McLeod.

Even with just one bird to account for, Keniston still had work to do—putting out food on the Green farm to lure the Heath Hen back, continuing to track down more apparent sightings and filing his daily reports that detailed such rumors, weather conditions and field notes. He re-

The world's last Heath Hen.

membered always that he must telegraph Boston should a second Heath Hen appear. No such bird did, so the rewards for sighting more Heath Hens—especially a female—went unclaimed.

With bad weather continuing into early March, hundreds of Vine-yarders contracted flu and pneumonia in early 1929, with one death a day on average. Keniston worried less about the Heath Hen and more about his sick child and wife. In his letters, he complained that winds made the house uncomfortably cold. His convalescing wife felt anxious to leave, to get away from the isolation and the controversies. How long would the last bird live? How long would Keniston be keeper of a refuge on which the last Heath Hen hardly ventured?

Keniston looked ahead. Thanking Gross for a recent gift to his son, he began to plan for a visit of the entire Gross family, but worried about the well-off writer Thornton Burgess bringing his wife and daughter—"we dont [sic] live in a palace . . ." (Before a later impending visit, Keniston awkwardly had to request that his friends Burgess and Gross pay for their meals while staying with the Depression-racked family.)

Sometimes, as Keniston's wife worked on a photo album for Alfred and Edna Gross and their children, Allan wandered to the yard and faced west again to the meadow where the Heath Hens once had gathered. Nothing had worked. "When we could hear the booming on the hill of fifty heath hens, then it was worth while and very interesting," he wrote to the professor, whom he always addressed as "Dr. Gross," despite the closeness of their friendship. "Now I listen in vain and it is depressing to think that in this comparative short period that they have vanished."

Vanished but one, who reappeared in the spring at the James Green farm, and, on an unforgettable day, did something no one had ever seen a Heath Hen do.

It was April 2, 1929, a cool Tuesday, the low having been below freezing and the afternoon high reaching barely into the 50s. Burgess and Gross huddled in the blind watching the lone bird mosey about Green's field. Then they caught their breath as the bird flew toward an oak tree—but not to the low branches for shelter. Instead, the Heath Hen flew to the very top of the tree, as if proclaiming his presence from that altitude might compensate for his loneliness.

Suddenly the bird—who had been silent so far that year—bobbed his neck, inflated the air sac, lifted his pinnae in the courtship V display, spread and pressed his wings against his body, lifted his tail and boomed. Far from the ground on which his companions once had walked, the world's last Heath Hen displayed his valiance and desire from on high, the sound of the moaning boom sliding downward to the drumming field where Gross and Burgess could hardly believe their eyes and ears.

The Heath Hen boomed and boomed from the top of the tree, and no kin answered. When a hawk flew over, its faint shadow rippling across branches and grasses, the Heath Hen pivoted down in flight and flew away to hide.

The end of the Heath Hen is a Janus-faced tale, with such tragic images, but also farcical moments, such as Keniston's morning in the blind on April 27. The Heath Hen, lately so visible, had returned to his skulking ways. "Mr. H.H." had not yet shown, Keniston reported to Gross. "D. the bird—! I know whats wrong he wont come here today as yesterday while here he didn't hear your seductive calls 'as one male to another' as Burgess says—Guess I will scatter some 'Rexal Orderlies' about the field—it is claimed they cause early operation every morning . . ." Whether Keniston resorted to baiting the Heath Hen with laxatives is a matter history cannot settle.

The weeks of spring heated to a dry summer and in that drought the local Heath Hen Committee finally disbanded, having done nothing to help the birds and having failed to oust William Adams from his directorship. The Martha's Vineyard Rod and Gun Club ("that bunch of nuts," Keniston privately called them) soon requested that the protections for this last Heath Hen be lifted in order to stock the island with Ring-necked Pheasants. "Of course I realize its [sic] all over," Keniston wrote to Gross, "but what nerve for those same chaps to ask him for pheasants so soon— why the poor heath hen hasn't got cold in his grave yet if he is dead??"

The state reaffirmed again its protection of this last bird, which would be left to live his days where he had always lived them, on the scrub-oak plains of Martha's Vineyard and the farm fields of Jimmy Green. After its autumn molt, the Heath Hen returned to feed at the farm in the fall of 1929. And poor "Mr. H.H." nearly found his grave in September when Harvard University's John Phillips, the president of the Massachusetts Fish and Game Association, almost ran over the bird with his car.

For Keniston the years of service—and insults and doubts—were enough. He resigned as warden on November 1, 1929, to become the manager of a former senator's farm while the state continued, indecisively, to consider whether to close the refuge altogether or to open it for hunters. Ultimately, officials decided to keep the refuge off-limits, until it was certain that the Heath Hen had gone extinct.

Just as the last bird persisted, the battles from years past persisted, too. Earlier in the year, an upset Alfred Gross had written to *The Wilson Bulletin*, the ornithological journal; it had published inaccurate statements about the history of the Vineyard Heath Hens. Gross summarized the matter of the state's refusal to allow Edward McLeod as much leeway in shooting as he wanted, saying that Arthur Clark's purpose was "to embarrass the work of the State Department for purely political reasons. This warden was killing all kinds of birds which had no possible relation to the protection of the Heath Hen." In his strongest language yet, Gross characterized Clark's articles as "a huge joke" and "malicious outrages."

Arthur Clark retreated from public view soon after this retort, but others were not through with controversy. For the next two years, Clark's allies launched a relentless letter-writing and publicity campaign to find a mate for the world's last Heath Hen. Newspapers and game departments in other states picked up the cause. Wisconsin offered to ship western-race Greater Prairie-Chickens to the island at no cost to Massachusetts. Strangers deluged Gross with letters urging him to support the importation of prairie-chickens to Martha's Vineyard.

The ornithologist and Massachusetts officials steadfastly refused, noting that previous introductions had proven unsuccessful. Even if a few new Greater Prairie-Chickens did manage to survive a release, the likelihood that the last Heath Hen would successfully mate seemed slim. "Mr. H.H." was probably sterile.

But, as if exasperated by years of public nagging and private confrontations, Gross—who had long opposed the importation of prairie-chickens—suddenly and inexplicably reversed his position at a meeting of the ever-agitated Martha's Vineyard Rod and Gun Club. The *Vineyard Gazette* hailed the professor's change of opinion, and all were certain that the state would finally give in as well.

Thornton Burgess, who had provided publicity for the Heath Hen's plight on his WBZ-Boston radio show, could not believe that his good friend had changed his mind:

> Gosh! Reckon it must be that paresis had set in before you got rid of them dern tonsils. It sure looks to me like softening of the brain . . .
>
> Here I am, getting kicked about as a sentimental ass by alleged sportsmen, who might have been expected to take pity on the poor old cuss for his lonesomeness, but do I? I do *not*. And here are you, a cold-blooded scientist, sorry the old bird has to sleep alone but primarily interested in the fact that he *is* alone and because of that fact the most famous bird in the world, boasting that it is the first time Science ever had had the opportunity to follow a species to the very end of the last specimen, turning around at the behest of a few weak-minded sentimentalists. . . .

Gross ate crow (a bird now being mistaken for Heath Hens, by the way) and acknowledged that the importation would confuse the record of what was indeed the world's most famous bird. He backpedaled quickly to his original position—no importation of Greater Prairie-Chickens.

A figurative feather-flying uproar of protest ensued—led in part by the *Vineyard Gazette*—even as the state continued to refuse the experiment. Well-meaning, if idiotic, suggestions to mate the last Heath Hen with a Ring-necked Pheasant or a Ruffed Grouse met with even terser rejection. As if to calm the fury, Gross began to emphasize in his articles and reports that the Heath Hen was indeed the last of the *eastern* race of the Greater Prairie-Chicken, not the last of the entire species.

As a symbol, however, the world's last Heath Hen remained singular. It galvanized interest in conservation. Articles on the bird appeared across the country, across the world. The last Heath Hen brought dramatically to life the dying of whole races of animals. Its lingering endurance not only made good copy but provided a powerful educational example. "The renown of the last specimen has spread remorse," said the *New York Herald Tribune*, "and the resolve to rescue other vanishing species."

The last Heath Hen—nicknamed "Booming Ben" by some—lived alone, silently moving between the scrub-oak plains and the Green farm as day and season dictated. He would gain no mate, no comrade. "It is truly remarkable," Gross wrote in 1930, "that this lone bird, subject to all the vicissitudes of the weather, to disease, and to natural enemies, has been able to live in solitude for such a long time." Everyone expected the bird to die each winter, to fail to show in the spring. State reports pronounced him dead, but he kept returning. Gross dubbed this Heath Hen "a sticker." His return attracted bird lovers, journalists and the merely curious to Green's meadow. The observers watched, waiting for a glimpse of a fat, well-preened Heath Hen, feeding on corn among the crows. In 1930, the annual census—hardly a census at all now, what with just one bird—took place between March 28 and April 4.

A year later, on April 1, 1931, Alfred Gross and Thornton Burgess sat in the blind on Green's farm, again watching the bird. It was 6:45 in the morning and a nor'easter blew unabated. Rain fell hard and steady, but the Heath Hen edged from the woods toward grain scattered by the blind. "Booming Ben" munched on the grain, then shivered his feathers and squatted down facing the wind. Would the ear of corn set as a lure in the middle of a trap entice the bird?

Both men knew, as Gross had told Burgess, that this was "ornithological history," another milestone. "Who would have thought it?" the professor had asked. "Last year I thought it would be our last. It is up to us to see the old bird through to the last."

Thirty minutes passed as Burgess and Gross whispered encouragement and goads at the rain-matted creature. Cold and miserable, the two men waited and waited. Then, suddenly, the Heath Hen got up and ambled into the confines of the low, iron-frame trap. He pecked at the corn, unfazed by a pole and the fishnetting high above him.

"It was an exciting moment inside of the blind; the least false move would mean failure," Gross wrote in his report to the state. "Promptly at a pre-arranged signal the trap was released." Either Gross or Burgess tugged on a rope attached to the pole, which collapsed the netting down upon the surprised bird.

They took pictures and movies of the world's last Heath Hen, which Gross pronounced healthy and "exceedingly strong." He estimated Booming Ben's age to be somewhere between seven and nine years—astonishingly old for a prairie-chicken, whose average adult life span in the wild is about one year. Using pliers, the men banded the bird, placing a tiny aluminum ring around the left leg with "407880" engraved in the metal. On the right leg went a copper band, engraved with "A-634024." This made mistaking the bird for a second Heath Hen less likely, and, if Ben were found dead, the bands ensured certain identification. Burgess and Gross each held the bird up, new anklets glinting, and they took more photographs.

Then they let the bird go. He scurried to the wet oaks.

Alfred Gross saw the Heath Hen for the last time on April 3 or April 4, 1931.

Neither Gross nor Burgess heard the bird boom that year. He had kept mute since flying to the treetop in 1929. But James Green told the *Vineyard Gazette* that shortly after the professor left, the Heath Hen did boom, just a little.

It was, apparently, the very last time.

The Heath Hen showed himself three times in May 1931, then vanished until February 9, 1932, when he reappeared at the Green homestead. The bird had weathered another turn of the year, but he now behaved oddly, avoiding the field where he had eaten so freely of corn and grain. Instead, he lurked at the edge of the scrub oaks, and, for nearly the next month, Green watched the newly wary bird.

March 11, 1932, was a cool day—in the 40s—and it was the last day that James Green ever saw the bird. Ornithologists now consider that to be the final confirmed sighting of the Heath Hen. Alfred Gross, in his privately printed autobiography, wrote that the last sighting was on April 6, 1932.

Gross had hoped to trap the Heath Hen again in order to inspect the conditions of the numbered bands, but he never got a chance. He also hoped to bring *caged* female Greater Prairie-Chickens to the island—to set out in the field as a living lure for the last Heath Hen. But Raymond Kenney, the new director of the Division of Fisheries and Game, apparently refused. It was moot. Gross and his students never found the bird.

On the last day of Gross's visit, however, two Vineyarders claimed they saw the bird up close, according to the newspaper:

HEATH HEN BEGS BREAD WHILE STUDENTS STALK

West Tisbury, April 2 [1932] (Special)—While a Bowdoin college professor and three student assistants are searching the islands, high and low, for a glimpse of the world's last heath hen, the bird is making friends with motorists at the old Indian cemetery here, for the sake of a few bread crumbs. Orrin A. Gardner of this town advised today that he and Anthony Campbell fed the famous heath hen at the Indian cemetery on Tuesday and that the bird came so close to the two men that they could have touched it.

Gardner told the *Vineyard Gazette* that he and Campbell could even see the Heath Hen's identification rings.

In 1932, James Green received his $5 for care of the blind and use of his field, though one suspects he would have traded the cash for later, certain sightings of a bird he'd come to love. Gross had given Green several self-addressed stamped envelopes to use for reports of any Heath Hen sightings. Green used only one envelope—for a note of March 13, 1933, telling the professor the bird had not reappeared.

After Gross returned to Bowdoin College he received a morbid note from John C. Phillips, a note that demonstrates just how far Phillips had come: A man who once wanted to save the bird felt differently now. Phillips told Gross, "I am afraid if the Heath Hen does not die pretty soon, we will all be at loggerheads. I hoped he would die this winter and I believe he has not been seen recently. That is good news." Phillips would soon be secure in his relief at the Heath Hen's demise.

Reports of sightings continued into July when someone saw what he called a "somewhat ragged-looking" Heath Hen. These summer sightings were never confirmed, and Gross's subsequent view in the published literature was that all later reports of the bird, extending into 1933, were of female Ring-necked Pheasants. Ludicrously, reported sightings of "Heath Hens" occurred as late as the 1960s.

No one ever found the remains of Booming Ben. But in 1997, Alita Prada, of Locke Mills, Maine, came forward to say that in 1932, her mother and a friend named Estenna Norton had hit the bird while Norton was driving on a foggy Vineyard night. The two women were so mortified that one of them flung the corpse as far as she could into the woods. Fearing prosecution, they never said a word about it except to relatives, years later. Of course, it's quite possible that the bird they killed was a Ruffed Grouse or a Ring-necked Pheasant and, Prada tells me, the two women never mentioned the identification bands.

Not long after it was clear the Heath Hen had truly gone, Laurence Fletcher, the secretary of the Federation of New England Bird Clubs, wrote to Gross with history on his mind:

> I picture just as you do that one hundred years from now some young or old ornithologists requesting Heath Hen records, will pour [sic] over them and remark what curious old cusses we were. Isn't it unfortunate, however, that there always has to be a political wrangle and unpleasant row attached to [each] movement of the Heath Hen story . . .

The *Vineyard Gazette* also addressed history on April 21, 1933, with a boxed headline on page one that read "Farewell to the Heath Hen," a long obituary on the creature and a moving editorial.

So, finally, the story had concluded. "Never in the history of ornithology," wrote Alfred Gross, "has a species been watched in its normal environment down to the very last individual."

What is seen today is nothing to what may be seen tomorrow . . .
—Henry Beetle Hough, *Birds of Martha's Vineyard*

The key actors in the drama on Martha's Vineyard went on with their lives. Allan Keniston stayed on the island and later managed the state forest; he died on December 12, 1979. Thornton Burgess continued to write his syndicated "Bedtime Story" feature for newspapers until 1960; he died five years later. Alfred Gross returned to teaching and to other research projects until he retired in 1953; he died in 1970. Gross has an island, a jungle trail and a subspecies of the Clapper Rail named in his honor; during his retirement, he traveled the world, visiting many places, including Mauritius, former home of the extinct Dodo. James Green continued to farm; he died in 1938.

But what of the protagonist? Like the phoenix, the Heath Hen may return. What might follow its extinction is—in a new century—a *new* Heath Hen. To understand how this resurrection is possible, we need first to understand the status of habitat management on Martha's Vineyard and to understand that, we need to meet a man named Tom Chase, who, when I visit him, is having a very bad day.

He's on the phone in his Nature Conservancy office off Lambert's Cove Road. Chase, a paleoecologist who coordinates the protection and management of 1,700 acres of Conservancy property on the island, grimly nods and shakes his head. He clutches the phone. He's just been told that the sparrows are missing. Of the two populations of the locally rare Grasshopper Sparrow, one group has not returned this year to the

Katama Plains on the southeast side of the island; a handful of the sparrows have failed to appear, leaving only eight or nine for the entire island.
This summer, 1998, no males sing their buzzy songs from the tops of
weeds or posts by the Katama Plains.

Why this disappearance? "The Grasshopper Sparrow responds best to
a two- to three-year burn cycle," Chase tells me after finishing his call.
"But one of the management partners resisted burning every two or
three years. So we're hoping that burning now will allow the area to be
repopulated by the sparrows around 2000." He pauses and looks intense.
"I knew I should have insisted more strongly on the frequency of burns."

For Tom Chase, fire isn't just a natural phenomenon or even a management tool. Fire is a recompense. "This all gets to how you view what
it is you call home," Chase says. "Humans are a forgetful species." We've
forgotten how to let the land itself—and the lives of animals—shape our
knowledge, expectations and desires. If we don't know the Grasshopper
Sparrow, for example, we'll never wonder about its needs. If we don't
wonder about its needs, they will go unmet, since humans control the
short-term ecological fate of habitats.

This perspective seems especially trenchant on Martha's Vineyard,
where, according to the *Boston Globe*, "More acres have been developed
since 1971 than in the preceding 330 years since Europeans arrived." Expensive, quaint houses, cottages and mansions vie for limited space, like
static versions of the traffic jams each summer brings, with throngs of vacationers navigating narrow streets in shiny Cherokees and svelte
Porsches. The winter population of the Vineyard numbers about 12,000;
in the summer, that swells to more than 70,000. Some of those tourists
shop at a fabric boutique called The Heath Hen Store, the only visible reminder of the former presence of that bird.

For all the elegance and specificity of his speech, Chase—in his blue
shorts, topsiders and short-sleeve shirt—could be taken as just another
tourist. He's not. He grew up on the Vineyard. He knows what's missing.

"My parents had the opportunity to hear [Heath Hens] booming. To me, the sand plains are silent."

If, as Chase suggests, "biodiversity is character"—the foundation of community—then the cultivation of biodiversity and the restoration of eradicated ecosystems and ecological processes quickly assumes a paramount importance. Fire belongs on the island and—if properly managed—it won't threaten the safety or property of residents and visitors.

Coming from Kansas, where every spring ranchers set hillsides of tallgrass prairie ablaze, I find the notion of controlled burns familiar. April days in my backyard, where I watch occasional migrating Hermit Thrushes and Tennessee Warblers, I breathe in the acrid scent of range fires just beyond the town's boundaries.

On the Vineyard, however, there isn't a strong tradition of managing fire as a way to create (or recreate) ecological niches for a variety of plants and animals that have come to depend on the scouring effects of flames. Nonetheless, longtime island residents and year-round locals seem to favor the burns. Off-islanders—the summer people—need more education about the benefits of fire. In an effort to reach that audience, The Nature Conservancy burned an area just off the Edgartown–West Tisbury Road, now a busy thoroughfare across the island. Informational signs explain that the blackened trees and earth aren't destroyed, but changed. From blazing star—a flower—to Short-eared Owls, a variety of island plants and animals need fire if they are to thrive. The Conservancy burned some 400 acres on the island in 1999.

When I toured the old Heath Hen reservation with State Forest Manager John Varkonda, he noted that an open grassland through which we were walking was one of the few left on Martha's Vineyard. He pointed at the ground. "This is incredibly dry soil," he said, "very prone to fire." In one of the ironies of the Heath Hen story, the dry land of the refuge, which became part of the state forest, sits atop the aquifer that supplies the Vineyard with drinking water, thus protecting the ground water from development and pollution. The central portion of the island is so dry that many years ago, according to the *Vineyard Gazette*, "Customers were sometimes given house lots [there] if they purchased a pair of coveralls or a jackknife at Island stores. Most of these giveaway lots went unclaimed because registering the deed for two dollars cost more than the land was worth."

Gesturing at the meadow, Varkonda recalled how, after the last prescribed burn in the state forest, in 1996, butterfly weed returned, as did an orchid, *Spiranthes vernalis*, spring or early ladies' tresses. The flower had not been seen on the island in more than a decade. Newly growing, waist-high scrub oak borders the meadow three years after it burned, while on the far side dying red pines edge against the grass and flowers. We looked at daisies, primroses, yarrows, black-eyed susans and even a patch of familiar prairie grasses, big and little bluestem, Indian grass, switchgrass. I sampled the blueberries at my shins and felt the pointed prickle of scrub-oak leaves—Heath Hen habitat—and asked Varkonda just how tough scrub oak really is.

The gregarious forester, dressed in khaki shorts, white tennis shoes and a green polo shirt, looked at me through his dark sunglasses and smiled. "In '96 we had flames that were 50, 60 feet high," he said. The fire actually melted some of the equipment used to monitor the burn. And the scrub oaks? Their root system can survive a fire as hot as 2,000 degrees.

The evidence touched my knees—a thicket of many-branched *Quercus ilicifolia*, dull-green leaves looking shiny in the summer afternoon sun. The scrub oak prospers with fire and, as if grateful for its life, provides others with a bounty. Several kinds of moth feed on scrub oak; in fact, nearly 400 species depend on the tree, though not all of them live on the Vineyard.

The return of fire also means the clearing of many of those ailing evergreens that now inhabit nearly a quarter of the 5,200-acre state forest. Planted in an effort to create a timber industry on the island, the evergreens—red and white pines and various spruces—have done poorly, succumbing to disease and the inappropriateness of soil and climate. Still, the exotic trees have managed to crowd out a lot of native plants. A 1999 Harvard University study recommended cutting and logging of 1,100 acres as part of the sand-plains/scrub-oak restoration effort. Discussions continue among state agencies and island conservation groups concerning funding and the pace of clearing through logging and controlled burns. One worry is that the clearing of mature trees could leave vulnerable patches inviting to poison ivy and other unwelcome competitors with native plants. Another factor is that the Harvard study has reopened a debate about whether the outwash plain was more scrubby or more grassy in decades past, an argument with implications for how—and what—to restore, once the sickly trees are gone.

Tom Chase and biologist John Toepfer aren't waiting for the state to implement a new forest-management plan. They are developing a proposal that will mean the return of a sound silenced on the island for three quarters of a century: the sound of courtship booming. Chase and Toepfer want to bring western-race Greater Prairie-Chickens to the island and establish a viable population of the birds, as a high-visibility consequence of the restoration of fire and scrub oak. Toepfer exudes confidence. He has no doubt that it can be done, even if, as the project gets under way, uncertainties remain.

Genetic tests of Heath Hen specimens and Greater Prairie-Chickens will determine their biological closeness. If the eastern and western races are genetically similar, and Toepfer believes they are, then it was their "culture" that differed, their behavioral adaptations to different habitat types. In that case, it is the *place* that will "make" Heath Hens out of Greater Prairie-Chickens. Place and time. "If we introduce Greater Prairie-Chickens to Martha's Vineyard," Chase reasons, "and they begin to adapt, and breed, maybe add 3,000 years, and then we'll have Heath Hens."

Whenever the first release takes place—perhaps in a couple of years—adult and juvenile birds will be placed in an enclosed open range, fitted with radio transmitters collared around their necks, and given all the time they need to learn the habitat, which likely will be land owned by The Nature Conservancy.

The chickens will be released when they are molting feathers—in late summer and fall—so they will avoid flying, which is painful during a molt. This will encourage them to get to know the unfamiliar ground, the new plants and trees. Toepfer says the island could support up to 25 to 30 cocks right now, but that isolated populations with fewer than 100 birds would have to be supplemented with periodic introductions of new birds from other flocks, in order to ensure genetic diversity. Probably the first group of Vineyard chickens would initially need about 1,000 to 1,500 acres of scrub-oak and grassland habitat, which will require between 3 and 10 prescribed burns to create. The entire project could take between a decade and a century to culminate in free-roaming flocks of proto–Heath Hens.

For Heath Hens truly to return, however, these adapting Greater Prairie-Chickens will need more room. The extinction of the original Heath Hens on the Vineyard underscores how vulnerable island populations can be—whether that island is a real one or a virtual one of appropriate habitat surrounded by monocultural crops or urban sprawl. Isolated populations of animals are more prone to extermination. To flourish, the proto–Heath Hens will need to repopulate areas on the American mainland. It is possible that a few Vineyard chickens someday would fly across the water and find mainland habitat on their own. One of the interesting conclusions Toepfer has drawn from his recent research for the Wisconsin-based Society of Tympanuchus Cupido Pinnatus is that cocks stay close to their birth sites while hens disperse as far as 25 to 40 miles away. But the hens won't disperse if only one lek is available. They'll stay near that lek, which results in the very problem that helped doom the Heath Hens on the Vineyard: inbreeding.

But will a population—or populations—of transplanted chickens reproduce sufficiently in the wild to offset annual mortality? "That's the big question," Toepfer acknowledges, "and it can't be answered until you have birds *there.*" The highly endangered Attwater's Prairie-Chicken on its public refuge near Eagle Lake, Texas, depends on periodic releases of captive-bred stock to maintain its population. The same might be needed, at first at least, with any Vineyard proto–Heath Hens. Because the mortality rate and the fledgling survival rate—on average—are both 50 percent, changing the factors that affect both can have huge population consequences.

There are other questions. "Can the introduced birds make it on natural vegetation?" Toepfer asks. Another concern is the high number of ground predators on the island, predators that could snatch a few chicken eggs for dinner. (A new kind of nesting fence may work to deter such raids.) Toepfer also suggests that hawks will be a major enemy of the new Vineyard chickens, since more than 80 percent of all prairie-chicken mortality is attributable to predation. Diseases and parasites will need further study. Managing the number and size of prescribed burns also will be complicated, given that such activity is a matter not only of biology but of administrative coordination among private and public conservationists, not to mention fire departments. Clear lines of management authority and cooperation must be in place for fire safety and for the management of the entire project.

Another hurdle has nothing to do with biology and everything to do with public relations. Already on the Vineyard the endangered Piping Plover receives human help. Needing sand dunes for nesting, the plover utilizes fenced-off areas for rearing young. The public appears to support the plover restoration effort, no doubt in part because of the inherent cuteness of the tiny birds. (On Little Beach, I watched four plovers race along the wet sand, one of them sounding for a minute, I swear, like a tiny cackling prairie-chicken trapped inside a shorebird's body.) Will islanders and summer people embrace the return of fire and the return of proto–Heath Hens? The ultimate success of the project may depend on it.

The Heath Hen's southern sibling, the Attwater's Prairie-Chicken, is living out its own version of the Heath Hen's story and suffers from a lack of public interest. A century ago Attwater's Prairie-Chickens numbered about one million birds along the coastal prairies of Texas and Louisiana. Now less than one percent of the original 6 million acres of such habitat remains. Overhunting and, more recently, habitat destruction have combined to put the Attwater's on the brink of extinction. Biologists working with this critically endangered bird had wanted to add, each year, five successfully reproductive females so that, by the year 2000, the flock would number 5,000 individuals. Biologists remain 4,954 birds shy of their stated management goal. Preliminary census figures from spring 1999 revealed a total wild population of about 46 birds, while the captive-breeding flock numbers 100.

Biologists are hampered by a lack of genetic diversity that weakens the birds over time, as random mutations accumulate, in a process called genetic drift. It's a tremendous problem in small, inbred populations. The mutations make the chickens more vulnerable to infertility and disease, as happened with the Heath Hen. In fact, researchers have discovered an AIDS-like virus in the Attwater's population and, recently, the disease was found in two Greater Prairie-Chickens in Oklahoma.

In Illinois, small patches of protected prairie echo with the spring booming of Greater Prairie-Chickens. The dwindling native Illinois population also had suffered from devastating genetic mutations that threatened eventually to kill off the birds, but the small flocks recovered because new genetic material from birds as far away as Kansas infused new life into the gene pool. The Greater Prairie-Chickens of Illinois—intensively studied and managed—continue to survive, albeit with a population numbering only in the dozens.

Greater Prairie-Chickens in Kansas number in the tens of thousands, a large population seemingly immune to threats. Even though the population trend for Greaters in Kansas was up in 1999, worry continues over range fires now set annually instead of every three years because those fires remove nesting cover. Concern mounts over an exotic plant, *Sericea lespedeza* or Chinese bush clover, which threatens to choke out native tallgrasses in the Flint Hills. Even urban sprawl now affects prairie-chickens in Kansas. Recently, a coalition of farmers, retirees and environmentalists in Manhattan, Kansas, successfully fought a county proposal to build an unneeded "arterial" road that would have destroyed a handful of prairie-chicken leks—only to have part of one lek destroyed by a new water tower meant to service a huge golf course and housing development. My friend Dave Rintoul, a biologist at Kansas State University, reported in April 1999 that he watched two male prairie-chickens displaying at that lek—next to a backhoe.

Protecting prairie-chickens—wherever they are—requires public education, intense political lobbying, sound biological management and preservation and restoration of habitat. On Martha's Vineyard the return of fire will lead to the recovery of scrub-oak habitat—and that should lead to the return of prairie-chickens, the eventual rebirth of the Heath Hen.

Perhaps someday a future ornithologist will hold a bird that has become so acclimated to the island to have earned the name *Heath Hen*. She will affix metal identification bands to each leg, then let the bird go—not as the last of a race but as part of a living flock. She will understand: Contemplation that does not foster action is no better than a broken wing.

The Passenger Pigeon

Overleaf: Passenger Pigeons. Note the netting operation in the background. From Thomas Nuttall's 1840 ornithology.

The Dark Beneath Their Wings

> [T]hese birds . . . have communicated to them by some means
> unknown to us, a knowledge of distant places . . .
>
> —Chief Simon Pokagon of the Potawatomi

IN A VOLUME OF HIS *AMERICAN ORNITHOLOGY*, PIONEERING NATURALIST
Alexander Wilson described a flock of Passenger Pigeons that he had witnessed in the early 1800s as the birds flew between Kentucky and Indiana.
The flock, Wilson estimated, numbered 2,230,272,000 birds. "An almost
inconceivable multitude," he wrote, "and yet probably far below the actual amount." The multitude spanned a mile wide and extended for some
240 miles, consisting of no fewer than three pigeons per cubic yard of sky.

Mathematicians and physicists perhaps can visualize the *number*, but
for years I struggled. Just what was a flock of more than 2.2 *billion* pigeons like? I needed metaphor. I needed to make the swarm linear. My
pocket calculator—good for figuring gas mileage—fritzed as I attempted
the equations. So I called on two friends with better calculators. What I
wanted to know was this: If the birds had flown single file, beak to tail,
how long would the line have stretched?

Assuming each pigeon was about 16 inches long, a line of
2,230,272,000 Passenger Pigeons would have equaled 35 billion inches, or
3 billion feet. That's 563,200 *miles* of Passenger Pigeons. In other words,
if Wilson's flock had flown beak to tail in a single file the birds would have
stretched around the earth's equatorial circumference 22.6 *times*.

Not to be confused with message-bearing "carrier pigeons"—those trained, domesticated birds so useful in war—Passenger Pigeons were wild creatures, prodigious and unequaled. This species once comprised 25 to 40 percent of the total land-bird population of what would become the United States. Historians and biologists have estimated that 3 to 5 billion Passenger Pigeons populated eastern and central North America at the time of the European conquest. The Passenger Pigeon was the most abundant land bird on the planet. The next time you see an American Robin, imagine 50 Passenger Pigeons in its stead; that was the ratio between the two during colonial times.

Jacques Cartier, the first European to write about the pigeons, did so on July 1, 1534, having seen flocks on what is now Prince Edward Island. Champlain saw them at Kennebunkport, Maine, in 1605. De Soto. Marquette. Sir Walter Raleigh. William Strachey. The pigeons awed them all. "So thicke that even they have shadowed the Skie from us," marveled one early account. "What it portends I know not," mused Thomas Dudley of Salem, Massachusetts, on March 28, 1631, after having witnessed a tenebrific flight of pigeons.

Flying as low as a few feet off the ground or as high as a quarter-mile, Passenger Pigeons moved in vast congregations that observers compared to squall lines, oval clouds, thick arms and waterfalls. Wilson saw how his flock flew in the shape of a river, then, suddenly, the birds moved into "an immense front." Flocks could contain pigeons on only a single level or be stacked in layers, with the birds flying loosely scattered or packed wing tip to wing tip. When bright sky showed through those multitudes, it must have glittered like a lantern signaling a frantic code, a frenzied semaphore.

With their powerful chests and long, quick-snapping wings, the pigeons flew an average of 60 miles per hour, for hours at a time. Sometimes the swift and seemingly endless flocks stretched across the entire dome of sky, so that wherever one looked, horizon or zenith or somewhere between, there flew the pigeons. They closed over the sky like an eyelid.

John James Audubon once watched a flight of Passenger Pigeons that, despite their speed, took three days to pass. "The dung fell in spots," he wrote, "not unlike melting flakes of snow." He compared the dark wrought by pigeons to the dark of an eclipse.

Thomas Nuttall called such flights of pigeons "a shower of sleet," a comparison made all the more interesting since sleet was one thing that drove pigeons from the sky to land. Mere rain typically would not. Fog also drove the birds down. In such conditions, the pigeons, especially young ones, would become confused and have to land. It didn't matter where—villages, churchyards, tavern roofs, barns, gardens. If the flock was over water, some birds alighted by chance on canoes or steamers; most drowned. This occurred not infrequently over the Great Lakes and less often along the Atlantic seaboard, even though pigeons tried to hug shorelines. They typically disliked crossing large bodies of water. One man saw drowned pigeons washed up on shore, "fastened together by their feet, looking like ropes of onions." The exhaustion of long migrations also drove pigeons to land recklessly in trees and shrubs, so wearied that a person could approach slowly, then take the birds by hand.

When he was a boy in Ohio, in 1876, Samuel Wharram daily fed wheat to a pair of pigeons overwintering on his family's snowy property. "They became so tame that when they saw me coming they would fly to me and often alight on my head or on the cup that I was holding. I often even picked them up."

In contrast to the quietude of Wharram's encounters, the flocks of millions roared like thunder, like trains, like tornados. (In fact, flocks descending to the ground sometimes looked like funnels.) One might get an idea of the noise by tuning a stereo to radio static, then gradually turning up the volume until the walls shake.

Passenger Pigeons were as beautiful as they were numerous. Their blue-gray heads, backs, and wings—which were spotted with small black patches—inspired Henry David Thoreau to say that this "dry slate color, like weather-stained wood . . . [was a] fit color for this aerial traveller, a more subdued and earthy blue than the sky, as its field (or path) is be-tween the sky and the earth." The pigeons' necks shimmered purple, gold, yellow and green, as if the feathers had been sprinkled with a metallic rainbow dust. "The reflections from their necks were very beautiful," Thoreau wrote. "They made me think of shells cast up on a beach." Shaped like Mourning Doves but larger, male Passenger Pigeons pre-sented thick, vinaceous chests that faded to white on their bellies. Their bills were black, though one pigeon specimen I saw had a bill the color of oxblood. The legs and feet were "lake-red," one ornithologist concluded, or, perhaps, "pinkish-red." Adults molted their feathers once a year, in late summer, and young pigeons acquired their adult colors by the spring fol-lowing their birth. Females had duller plumage but both sexes sported long tails and eyes so red and resolute they startled anyone who saw them.

Those red eyes—10 billion of them—would scan for food as the flocks flew above the vast deciduous forests of eastern and midwestern North America. The pigeons had evolved a reliance on mast (nuts) produced by oak, chestnut and, especially, beech trees. Because these trees produced mast in differing quantities at different times in different places, the pi-geons had to search nomadically for this capricious, nutritious food. (The species's scientific name *Ectopistes migratorius* roughly translates as "wandering wanderer.") Mast fruitings produced overwhelming num-bers of nuts—the same strategy, biologists point out, that the pigeons themselves relied upon: Overwhelm competitors with sheer numbers. Beech mast appeared every other year in different locales, but huge crops of nuts occurred just once every three to seven years.

Probably the birds could not actually see nuts from high altitudes; they looked first for appropriate trees, according to A. W. Schorger, author of the definitive natural history of the species. The pigeons would sweep down for closer inspection only if they found a promising expanse of mast-bearing forest, and, Schorger has suggested they likely remembered the locations of good mast crops from previous seasons.

The hunger of Passenger Pigeons was epic. If each pigeon in the flock that Wilson described ate about an eighth of a pint of mast each day, the total amount consumed in a year would have filled, according to one estimate, "a warehouse 100 feet high, 100 feet wide, and 25 miles long." Gleaning from the ground or winnowing on branches to pull mast free, the pigeons gorged themselves, able to open their bills an inch wide to accommodate the nuts. So crammed with food awaiting digestion, pigeons' crops often swelled to the size of an orange or larger. One pigeon was found to have in its crop 17 acorns; another had a half-pint of beechnuts; yet another had 104 corn kernels, 14 beechnuts and 22 maple seeds. (A crop that full could rupture if the pigeon happened to fall to the ground or, as sometimes occurred, flew into the side of a barn or some other structure.) After eating, pigeons would perch and rest, puffing their feathers up and placing their beaks upon their breasts.

One naturalist compared the motion of a feeding flock of pigeons—with birds in the rear flying constantly to the front in order to secure mast—to "the appearance of a rolling cylinder having a diameter of about fifty yards, its interior filled with flying leaves and grass . . ." Not even snow deterred the determined foraging of pigeons, who, picking and sweeping with their bills, would leave in their wake a messy mix of plant matter, soil and snow stretching for miles.

The rolling waves of red-and-blue-gray birds that moved across forest floors and clearings seemed almost to vacuum the mast until it disappeared. This, no doubt, put enormous pressures on other nut-eating animals, such as squirrels and Blue Jays. The coming of Passenger Pigeons also meant dread to farmers, who worried that their hogs would have too little mast on which to feed and therefore could starve.

In poor mast years, the pigeons typically fed on blueberries, huckleberries, pokeberries and other such fruits in the Great Lakes region, New England and Canada. The birds ate seeds and insects, such as grasshoppers and locusts. They kept caterpillars in check and relished earthworms and snails. Like the Carolina Parakeet, the Passenger Pigeon loved salt and ate grit and gravel to aid digestion. And the birds had the fetching habit of puking when they found something else they would rather eat. Pigeons twittered and squeaked while feeding, a sound multiplied by thousands and millions to become deafening.

The Passenger Pigeon made sounds that have been compared to the croaking of frogs, the creaking of trees and the tinkling of sleigh bells. During the breeding season, pigeons called a weak *coo-coo-coo-coo;* during copulation, they growled softly. A typical call for gaining attention or intimidating another bird was a harsh *kee-kee-kee-kee,* which started loudly and grew softer, with the pigeon holding its head back, opening its bill, spreading its wings up and out, then, with neck stretched, uttering a concluding *coo.* Pigeons calling from the sky to pigeons on the ground, and vice versa, repeated a shrill *tweet.* When alarmed, the birds took on an erect, alert posture, pressed their wings against their bodies, nodded their heads in a slow-to-quick circular movement and called a danger note that sounded "like a laugh made with a child's trumpet," according to naturalist C. J. Maynard.

One could produce a fair simulation of the pigeons' creaking sounds—prating—by inserting a silk band between two wooden blocks and holding the silk "between the lips and teeth [then blowing as if upon] . . . a blade of grass between the thumbs," explained a person familiar with the technology. Or one could buzz, through tightened lips, two high-pitched monotones for a few seconds, then a longer, higher-pitched monotone, followed quickly by two brief monotones descending an octave. When I tried this in my yard I succeeded in frightening two Mourning Doves and

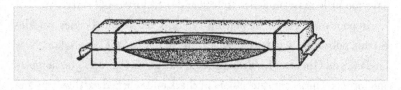

The pigeon call.

a Northern Cardinal and attracted the interest only of a single baffled squirrel that chuffed its disapproval.

Passenger Pigeons bred from April to June, principally in the northeast United States, the Midwest and north into Canada, areas often still covered with snow when the birds first arrived each spring. This behavior apparently ensured that the pigeons got first dibs on nuts left beneath the snow from the previous autumn. Although cold rarely bothered Passenger Pigeons, snowstorms could force northbound flocks to turn around, and an especially hard cold snap or a snowstorm could kill the birds once they were on their breeding grounds. Hard snows sometimes sent desperately hungry pigeons to seek food among farms, gardens, yards and villages. Snows falling late during incubation could cause the birds to abandon their nests and attempt a second nesting elsewhere, if possible, or give up breeding for that year.

A portion of the Passenger Pigeon population did breed in isolated pairs and small groups even in the heyday of the species, biologists have pointed out. But most of the population gathered in nesting colonies whose *average* size stretched over 31 square miles. Staggeringly huge aggregations of pigeons could occur. An 1878 nesting near Petoskey, Michigan, covered more than 200 square miles. The largest known nesting, an 1871 colony in Wisconsin, scattered across *850 square miles* and may have included nearly all the Passenger Pigeons alive in North America at the time, some 135 million birds. Not every square mile was chockfull of pigeons, though, as the birds concentrated in certain sites and did not occupy others within that territory. Thus, these large colonies were more like "a chain of nestings," wrote Schorger.

Nesting sites changed annually, but typically one large nesting occurred in New York or Pennsylvania while another took place in the Midwest or upper Midwest, usually Wisconsin or Michigan. The nestings usually began along streams and creeks, then expanded to higher terrain. Biologists are still not sure why some areas were selected for nesting sites when other locations were closer to supplies of mast.

All stages of breeding occurred in highly synchronous fashion—courtship, nest-building, egg-laying, incubation, hatching, feeding and, even, abandonment of the young. That is, all the pigeons performed the same activity within the same time period.

When the birds had selected a nesting site, they would flap their wings to break their flight, descending en masse, and would dip down like a cracking whip before rising up into the trees. The pigeons piled onto branches to begin their courtship rituals, but so many birds crowded together that the branches sometimes crashed to the forest understory.

In courtship, males jealousy guarded their mates and vociferously voiced anger at the approach of rivals. Courtship involved billing and sidling and necking (a male putting his neck over his chosen hen's neck). Amorous males and females sometimes rushed together and put their wings across the other in a kind of hug, a practice that terrorized pigeons of other species with whom, in captivity, Passenger Pigeons sometimes tried to mate. In the wild, Passenger Pigeons took three days to attend to courtship and copulation.

Next came the gathering of twigs, which males brought to the females one at a time, with much cooing and touching. A male would perch on the back of a female as he passed a twig from his bill to hers. All the females in the colony spent the same three days building their nests, and the pigeons used many kinds of trees for their temporary homes. A single tree could contain many tens of nests, sometimes approaching or exceeding 100 per tree, spaced from a few feet off the ground to as high as 50 feet or more. Throughout the ample literature on the Passenger Pigeon, writer after writer noted how flimsy the nests appeared, yet Schorger claimed that they were better constructed and sturdier than most believed. When males exhausted the supply of forest twigs, the birds might filch from each other or rob haystacks.

Then the female pigeons each laid one egg—synchronously all on the same day—and the adults incubated it for 13 days. The males relieved the females of incubation duty from about 10 A.M. to 3 P.M. each day, so the hens could fly off to feed. These shift changes—again, en masse—looked so spectacular that one writer said it seemed as if "earth and forest [had] been converted to feathers." From nestings or roosts, the pigeons would roam many tens of miles, and more, in order to find food. The male birds also had to fly to and from their nightly roost, which was separate from the nesting colony.

The hatchlings, blind and naked at birth, soon acquired a yellowish down. They would spend a total of two weeks in the nest, and the parents again took turns tending to the young, with the males relieving the females around midday. As a young bird grew, both parents would begin to leave the nestling briefly alone.

Adults fed their nestlings, called squabs, with "pigeon milk," until the young were able to eat softened food. Pigeon milk—which was rich in proteins, fat and salt, though lacking sugar—actually looked more like a white curd than milk. The young bird placed its bill in the adult bird's open beak to stimulate the flow of pigeon milk from the parent's crop. The squab then opened its bill and supped greedily. Sometimes the milk spilled from the adults' beaks to form long white lines down the pigeons' chests. Despite the colonial nature of the nesting, parents would not feed another young bird unless their own squab died. In that case, adult birds became desperate to rid themselves of built-up milk. It could kill them if they didn't. If another pigeon, even another adult, placed its bill into the beak of the desperate bird, the built-up milk was released.

When the two weeks of tending squabs ended, a spectacular event ensued: the abandonment of the young. After the young had been gorged in a final feeding by the parents, all the adult birds—every single one—flew out from the forest, formed a giant flock, and left. The squabs, stunned and forlorn—and with crops nearly as big as the rest of their bodies—squalled for nearly a day. They fluttered down fecklessly from their nests and tumbled about on the ground. They spent three or four days foraging and fluttering, awkward with their short tails and ungainly plumpness. But soon they slimmed down and, like the adults, flew off all at once in their own massive flock.

After breeding, in late summer, adult and young pigeons would fly to southerly roost sites to spend the fall and winter. Mild weather would sometimes find them roosting farther north. Winter roost sites were traditional in some locales, depending on food availability. This meant that fidelity to a roost could last for hundreds of years, even if it was not occupied annually. After an absence of a few years, the pigeons would return to an old roost site and use it for a few winters until switching locales again. Roosts could be hundreds to thousands of acres in extent, ranging through canyons, piney woods, osage orange groves or, the pigeons' favorite, swamps. From time to time, roosts formed on human structures such as bridges or even railroad tracks. Like all matters relating to the Passenger Pigeon, these roosts embodied a grand, chaotic anima.

Consider, for example, how the roosting birds set out at dawn to feed and to drink water. Seen from a fair distance, these departures must have seemed like the eruption of ashes or smoke from a volcano. The roar of wings and the calls of birds could be heard up to three miles away, as if the earth and air rumbled with premonitions. The birds liked to drink from the shores or banks of rivers and lakes, but if a clear bank was not available, they might alight on water for a fraction of a second, wings partially spread, sip, then lift back up into the sky. In at least one reported instance, pigeons were so thirsty that they landed on each other's backs over water, driving their kin into the small lake and drowning thousands of them.

In the late afternoon and at dusk, the pigeons flew back to their roost. It often took hours before the entire flock had returned. An evening appearance at one Iowa roost provoked this metaphor: "Their continuous arrival resembled the pouring of a sheet of water over the incline of an apron in a dam across a stream." So many birds would congregate at a roost that late arrivals had to land on the backs of other birds and jostle for position. As in the nesting colonies, tree branches crashed down beneath the weight. "A roost was a bedlam," Schorger wrote. A stinky one at that. Dung at one roost piled two and three feet thick in some places. This could result in dead trees and soil made rich with guano, ensuring a fertile field for years to come, should a farmer decide to brave the stench and begin planting after the pigeons had gone. A more immediate benefit accrued to those who stalked the pigeons at both nesting colonies and roosts—namely, dinner. The pigeons' doings attracted a host of predators, including wolves, bobcats, minks, martens, foxes, bears and hawks.

Few sights could compare to the pursuit of Passenger Pigeons by hungry hawks. Responding with an instant burst of aerobatic maneuvering, the pigeons would accelerate their flight speed to about 70 miles per hour, and the Merlin or Sharp-shinned Hawk or Northern Goshawk or Peregrine Falcon got the chase of its life. Audubon said that hawk-chased flocks became "almost solid masses . . . [that] descended and swept close over the earth with inconceivable velocity, [and which] mounted perpendicularly so as to resemble a vast column, and, when high, were seen wheeling and twisting . . . [like] the coils of a gigantic serpent." Despite their speed and contorted pivots, the pigeons rarely collided in flight.

Native Americans also took careful note of pigeon roosts and nesting colonies. Indians killed Passenger Pigeons for feathers, food and fat (squab fat made an excellent butter). To gather the birds, Indians would use nets and poles, as well as blunt-nosed arrows that knocked squabs from their nests. But the hunts never took place at levels that diminished the species in any significant way. Tribes rarely killed nesting adults beyond what was needed for subsistence or trade. Some tribes would not kill nesting adults at all and threatened white settlers who did. "Under our manner of securing them," said Chief Pokagon, of the Potawatomi, in 1895, "they continued to increase."

O-me-me-wog to the Potawatomi and Omimi to the Algonquin— these names echoed sounds the pigeons made. To the Narragansett, the pigeon was Wuscowhan—wanderer. The Choctaw called the pigeon Putchee

Nashoba—lost dove. The Hurons believed the souls of the dead came back as pigeons, and the Seneca offered gifts of tobacco to the adult birds before taking any squabs. The Micmac cleverly turned the hunted pigeon into a hunter itself. In a Micmac legend, one of the stars between the Big Dipper and the Northern Crown represents a Passenger Pigeon, part of a covey of different birds forever pursuing a bear (the Dipper's bowl).

American settlers also had interesting beliefs about Passenger Pigeons. Cotton Mather stated that when the flocks disappeared after a roosting or nesting, they would "repair to some undiscovered satellite accompanying the earth at a near distance." The more observant Benjamin Smith Barton, author of a 1799 American ornithology, the first to be written by an American, noted that the birds sometimes overwintered in and near Philadelphia if the weather stayed temperate. Such mild winters led to spring plagues, and the pigeons, some thought, were responsible for the outbreaks. Barton suggested that a poet could take this subject and rhyme it into a fine poem. (Longfellow later took him up on that challenge in one part of *Evangeline*.) An 1834 guidebook to dreams claimed that "To dream you see pigeons flying, imparts hasty news of a very pleasant nature, and great success in undertakings." In a more practical, if misguided, vein, housewives would sometimes ask hunters for dead pigeons—males only—believing that, if buried among the flowers, the pigeons' red breasts forced into the blossoms a brighter color. Some settlers even used pigeon dung and gizzards as folk medicines.

One belief about the pigeons mattered most of all: that they would abide. "Wonderfully prolific," wrote an Ohio legislative committee of the Passenger Pigeon in 1857. "No ordinary destruction can lessen them or be missed from the myriads that are yearly produced."

Ordinary destruction often meant the killing of pigeons to protect crops, for the birds could quickly clean a pioneer's newly sown field of its seeds. This occurred only when mast was not available, but the problem seemed severe enough to prompt the invention of the first underground seed drill in 1860. Farmers used nets to capture pigeons, bells and scare-

"Shooting Wild Pigeons in Iowa," from Frank Leslie's
Illustrated Newspaper, *September 21, 1867.*

crows to scare them and guns, of course, to shoot them. Diligent Galusha A. Grow of Susquehanna County, Pennsylvania, spent his spring days in 1834 atop a barn roof rattling sticks to affright pigeons interested in his family's crop of corn. He would rise before sunup and take his meals on high. The business was serious, for a lost crop could mean financial ruin or starvation. The Plymouth colonists lost crops in 1643 because of pigeons, and the people went hungry. In Canada, a bishop tried to excommunicate the pigeons because of crop depredations there. But the extent of agricultural damage caused by the species is impossible to know.

The pigeons did not always bring ruin. Samuel Goodrich, in his 1800 report on the Connecticut town of Ridgefield, noted that pigeons once destroyed legions of worms threatening to ravage apple orchards. More important, though, these birds were edible, and so settlers immediately began to kill them. In 1648, enough pigeons were killed at Plymouth to stave off hunger after a bad crop year. Nearly two centuries later, one Massachusetts man who had been wounded at Bunker Hill and who had lived as "pigeoner" for 45 years declared: "I have no family to provide for, thank God, save my grandson, and he, when I die, shall have my blessing, my bible, and my gun—as long as there are Pigeons in the world, we shan't starve." After a day's work of luring and shooting pigeons, the old man, grizzled and satisfied, would go home and prepare some for the dinner table. He sold the rest at 50 cents per dozen and each pound of feathers for 12 and a half cents. (Over six weeks of shooting, he cleared $4 daily, enough income in the 1830s to support himself and his grandson for the entire year.)

When printing presses became available, cookbooks were soon published—and many of them included recipes for Passenger Pigeons. They "make a very nice pie in the same way as chickens," according to an 1857 cookbook. In 1895, popular nature writer Mabel Osgood Wright asserted, "Old housekeepers remember when, in New York and Boston every winter, carts loaded with these birds went from door to door and potted pigeon was a standard New England dish, alternating with roast beef, turkey, and sparerib." Passenger Pigeons were broiled, roasted, pickled, smoked, salted or covered with dripping fat. They were rinsed, boiled briefly, drained, then stuffed with charcoal as a preservative. In America's growing cities, the pigeons provided a rich hostess or restaurateur with an elegant dish for a fancy table. Advised the 1819 edition of Maria Eliza Rundell's *American Domestic Cookery:*

> Pick two very nice pigeons, and make them look as well as possible by singeing, washing, and cleaning the heads well. Leave the heads and the feet on, but the nails must be clipped close to the claws. Roast them of a very nice brown, and when done, put a little sprig of myrtle into the bill of each ... The head should be kept up, as if alive, by tying the neck with some thread, and the legs bent as if the pigeon sat upon them.

On the frontier the pigeons quickly became a staple food, whether served with salt pork, chestnut stuffing or a gravy.

In town squares and city markets, shoppers had the choice of buying recently shot adults, which might be lean and tough if they had been taken right after a long migration, or the wild squabs, which were universally esteemed as "very delicate eating," in the words of one market guide. Advised an 1831 housekeeper's manual, "The good flavour depends very much on their being cropped and drawn as soon as killed.— No other bird requires so much washing." If live pigeons were available, shoppers could opt to have the birds dispatched at the market or else take the caged birds home, where they might be fattened with a paste of honey, cumin and fried beans.

Into the early nineteenth century, settlers and Native Americans did not pose a very serious threat to the Passenger Pigeon, even though hunting for markets—not just for subsistence—was clearly under way. In May 1771, a single day at the Boston market saw 50,000 pigeons for sale. But it wasn't until the middle of the nineteenth century that pigeon hunting grew from levels of sustainable harvest and became a voracious and highly profitable commercial enterprise. The pigeon industry employed at its peak hundreds, if not thousands, of men who descended on nesting colonies and roosts to slaughter the birds. Pigeoners especially owed the frenzy of their industry to the development of the railroad. Biologists David Blockstein and Harrison Tordoff point out that America had only 23 miles of rail in 1830, but by 1860, 30,000 miles of track had been laid. In 1857, it took only 48 hours to get to New York from Chicago; 50 years before, it had taken over a month. Newly speedy travel enabled pigeoners to deliver their goods more quickly to market. The development of the telegraph only enhanced shipping efficiency, because this instant form of communication let pigeoners know exactly where their prey was flying or landing. The Passenger Pigeon, it turned out, did need protection because the destruction became extraordinary.

Hunters procured the pigeons by two methods, netting and shooting. To attract a flock of wild pigeons, men cleared moist, fertile soil of vegetation, then spread salt, grains (often soaked in whiskey to make the birds drunkenly pliant), saltpeter, sulphur, anise seed, anise seed oil and/or worms in various combinations kept secret by each trapper. Netters often used natural salt beds as their trapping grounds. From a bough house (a temporary blind made of cut shrubs and tree limbs) the netters lured flocks down with pigeons flying high at the end of a rope—"fliers"—and with live decoy birds closer to the ground—"stool pigeons."

Netters sewed the eyelids of the stool pigeons together for the duration of the trapping season and ministered to their care with the attentiveness that desire for profit focuses. They fed and watered the birds by hand. After all, it was hard to train stool pigeons. Good ones could cost as much as $25 a bird. In the field, stool pigeons wore little booties of buckskin that were attached by a cord to a perch made of a flexible wood. The perch, with the pigeon on it, would be raised three or four feet by means of a rope running to the bough house; when the rope was let go, the perch

"Netting Wild Pigeons in New England," from Frank Leslie's Illustrated Newspaper, *September 21, 1867. Note the stool pigeon just visible at the left side of the prepared bed.*

fell and the stool pigeon fluttered—a sight that wild pigeons took to be one of theirs just landing. Trappers also used stuffed or propped-up dead birds as decoys.

When netters sighted a flock, they would release the fliers and flutter the stool pigeons. The men sprung the net just as a wild flock was about to alight. The nets—attached to hickory poles and weighted with lead sinkers—quickly covered the frightened pigeons. Chirping and grunting in terror, the birds were thus forced to land. Crews plunked stones and weights along the periphery of the nets to keep the birds from escaping. Sometimes, though, netters let a few pigeons go if the mass was too large for the nets to contain. (Alternatively, nets were stretched between posts across known pigeon flight paths. At night roosts, men arrived with torches and stones to drive birds into such webs.)

One witness said that after a net sprung, "Confusion ten times confounded follows and such a fluttering of wings and straining to get free you never saw . . . The pigeons, frightened beyond any power to describe, put their heads through the meshes of the net to fly." This actually made the exhausting work of killing the pigeons easier: Netters walked across the undulating mass and broke the birds' skulls or necks. The netters used pincers to do this, which was speedy but bloody, or snapped the bones with their fingers. This could go on for long hours, leaving hands so sore the fingers couldn't move. Biting the heads to crush the skulls also worked, but here too one's mouth and teeth grew sore by day's end.

After a haul, netters had to clean their equipment because Passenger Pigeons, upon seeing feathers of their kind or any blood at all, became so wary and frightened that they would not land. Pigeoners often sold the gathered feathers, as 50 pigeons yielded a pound of bedding.

Trappers could capture hundreds or even more than a thousand pigeons with just a throw of one net. A single pigeoner could haul in, after a season, usually spring, several thousand to tens of thousands of pigeons.

Netters also made a profit from keeping captured pigeons alive. A Dr.

Isaac Voorheis caught 1,316 pigeons one season in 1880 or 1881, and earned a tidy $650 with only six throws of the net. He sent the live birds via schooner from Michigan to Chicago. While some pigeons were sold as birds for domestic keeping (Audubon took more than 300 pigeons with him to England to present as gifts), live birds usually ended up as Voorheis's birds did: as targets. Trapshooting pigeons began in about 1825 and became wildly popular, spawning an industry devoted to manufacturing special cages that released the birds, as well as patented mechani-

Grand Pigeon Match,

At the Golden Lion Inn, Yonge-Street.

A Grand Pigeon
SHOOTING MATCH

Will take place at *Sheppard's Inn*, as above, on **WEDNESDAY, 26th of** SEPTEMBER, instant. Upwards of **Three Hundred Pidgeons** are provided for the occasion, and it is purposed to give **Three Prizes** as follows :

For the Best Shot, a Prize of £10 —Second best do. £5.—Third do. do.—a Good Rifle ! ! !
☞ Shooting to commence at 11 o'Clock, before which Hour the Gentlemen wishing to participate in the Sport, will be required to enter their names, and to comply with such Regulations for the government of the Sport, as may be arranged amongst themselves after their arrival.

Dinner will be on the Table at 4 o'Clock.

YORK, 16th Sept. 1833.

[G. P. Bull, Printer, " Courier" Office, Market-House, York.

Advertisement of an 1833 pigeon shoot.

cal devices, such as wooden "cats" or "agitators," meant to scare the pigeons to fly. One estimate suggested that a quarter- to a half-million pigeons were netted annually between 1825 and 1880 strictly for use as contest targets.

Urban gentlemen vied for bragging rights at pigeon shoots. Perhaps the best trapshooter was Captain Adam H. Bogardus, who once dispatched 500 Passenger Pigeons in 528 minutes while loading his own gun. An 1881 Coney Island competition resulted in the slaughter of 20,000 birds. One trapshooter and organizer of such events recollected that he had personally killed nearly 30,000 pigeons over the course of his life. Even ornithologists participated in the shoots, including William Brewster, who, several years after one such contest, searched Michigan in 1888 to see if the pigeons there had gone extinct.

The matches themselves, despite the patina of sporting gaiety and refinement, involved much cruelty. Contest promoters had to injure or mutilate pigeons before placing them in a cage so that when they were released they sprung more sprightly into the air. One spectator saw many such birds "bleeding and crippled, writhing in agony . . ." An 1875 contest featured Passenger Pigeons and Dark-eyed Juncos, who, being reluctant to fly, were prompted to do so by having snowballs thrown at them. Pigeons bound for trapshooting contests were shipped in cramped cages that caused the birds to lose or soil feathers. And the pigeons arrived with dire thirst, so that when water finally was offered the mad rush to drink caused riots in which many birds died. These various brutalities provoked Henry Bergh, founder of the American Society for the Prevention of Cruelty to Animals, to lobby for seven years until New York banned trapshooting. Public protest eventually muzzled the sport by the late 1880s.

What never stopped, until the birds were gone, was the relentless hunting of pigeon flocks, whether in flight or at nesting colonies or at roosts. Burning sulphur, left in pots beneath pigeon-filled trees, could fumigate the birds until they tumbled down. Burning grass had a similar effect. Nests in birch trees were susceptible to hunters setting fire to the ragged, papery bark, which flared up and scared squabs to the ground. One could also poke the birds out with a pole.

William Brewster observed that Passenger Pigeons sometimes "gazed fixedly at the advancing sportsman with what seemed an expression of blended timidity and curiosity." Though the sound of a breaking branch often startled feeding pigeons, the birds lost their wariness, such as it was, once they were nesting.

Sticks and stones knocked pigeons from the air in midflight. If the birds were flying close, a man could reach out with his hand and grab one. Sometimes men set up poles of hickory, making them wave by pulling on an attached cord; the poles slammed many birds down. This technique worked best on young, inexperienced pigeons. In at least one instance, *whips* were used to kill flying pigeons. In early March 1860, George Baker, a Cleveland fireworks dealer, launched rockets into a flight of innumerable Passenger Pigeons, causing a "wild confusion."

At one pigeon hunt, in a forest, a witness described how crews cut down 1,500 acres of trees and "as fast as the trees fell, the half feathered and fluttering fledgelings [*sic*], scarcely able to fly, were seized or knocked down with sticks, their necks quickly wrung . . ." Another witness called such scenes "a pandemonium for a saturnalia of slaughter." Men standing in inches, or feet, of dung had to shout to be heard above the din of birds and guns.

One could even shoot without aiming into a passing flock or at perched birds and be assured of killing several or several dozen or even 200 to 300 pigeons with one squeeze of the trigger. Author Morris Schaff recalled how he and some farmhands fired at random into a roost at Bloody Run swamp, Ohio, and bagged six bushels of pigeons. It was a late autumn night, he recalled, a night of awful beauty, with "a fleet of heavy clouds . . . sailing across a full moon."

Shooting young birds often increased one's take, for the inexperienced pigeons flocked nearer each other when shot at, not unlike Carolina Parakeets.

Men sometimes constructed ladder-like latticeworks on which the birds would alight; using stool pigeons and fliers, the gunners attracted birds to land on the spars, which were easy to rake with gunfire. Hunters also designed slatted traps in the ground with decoys or, more simply, baited a trench. All one had to do was fire at the birds massed together.

"The passenger pigeon was shot so easily," Schorger wrote, "that up to the middle of the nineteenth century it was not even considered a game bird." So easily slaughtered was the pigeon that even its hunters grew tired of and queasy with the ceaseless roar of birds, the bone-rattling shots and the smell of dung and corpses. Still, a good gunner could dispatch several hundred before breakfast, which was money in the pocket. (A meal break gave men time to pour water down gun barrels—to cool the metal.)

To process the birds for shipping, hunters had to clip wings, pull tail feathers and, with squabs, snap off their heads and crops to slow down spoilage. Then workers packed the pigeons in ice, 25 to 35 dozen per barrel. The barrels made their way on trains bound for New York, Boston, Chicago and other cities.

The numbers of pigeons shipped are staggering. The 1851 Plattsburgh, New York, nesting ended with market shipments of 1.8 million pigeons. The Grand Rapids, Michigan, nesting in 1860 yielded only 200,000 pigeons shipped to market, but with the tremendous waste of carcasses and with locals gathering as many pigeons as they could for their own dinner tables, some 1 million birds died there. A pigeoner in Monroe County, Wisconsin, shipped 2 million pigeons to market in 1883. H. B. Roney, an East Saginaw, Michigan, professor who agitated on behalf of stronger protections for the species, wrote in the *Chicago Field* that just three 1875 nestings in Michigan had yielded 1,000 tons of squabs and 2.4 million live birds captured. One man reported to Roney that he had walked through a warehouse whose entire floor was three to four *feet* deep in dead pigeons.

Daily incomes for pigeoners had increased from $4 in the 1830s to $10, then $40, in the late 1870s and early 1880s. Dealers such as N. W. Judy of St. Louis, Henry Knapp of Utica, New York, H. T. Phillips of Detroit, and Holmes and Sears of Chicago did a considerable business.

When city markets became glutted with the pigeons, the price dropped, so hunters would leave thousands upon thousands of birds to rot on the ground where they might be eaten by farmers' hogs. In the mid-nineteenth century market prices could range from 12 cents per dozen to a dollar per dozen.

Samuel Wharram, the Ohio youngster, took pity on pigeons netted on his father's property. The boy hurled stones at the pigeons to try to drive them away. In 1834, a Louisville actor actually prepared a funeral for and buried a bushel of Passenger Pigeons. But Wharram and those like him remained a distinct minority. While some publications and sportsmen had announced support for protecting pigeons, especially the squabs, these sentiments were rarely acted upon.

In fact, the first laws related to Passenger Pigeons protected townspeople, not birds. Flights of pigeons brought crowds onto city streets, firing recklessly, and such dangerous behavior needed to be controlled. In York, Canada, an attempt to restrict shooting in town failed when the police joined the revelry of killing.

Laws protecting Passenger Pigeons were finally enacted in the late nineteenth century, most notably in Michigan and Wisconsin. But the statutes had been written with the help of the commercial pigeoners. This ensured plenty of ambiguity and rendered the laws largely ineffective. Both Michigan's law and Wisconsin's, for example, allowed *limitless* netting of pigeons at some distance beyond the nesting area. Hunting was banned only *within* the nesting colony, a rule most pigeoners ignored anyway. Absurdly, the Michigan statute protected only nestings *after* a never-defined "last" hatching. Though a few netters had been fined at the

Shelby, Michigan, nesting in 1876, officials rarely enforced the regulations, loose and ineffective though they were.

In a heroic attempt to both enforce Michigan's law (as much as was possible) and simultaneously highlight its weaknesses, H. B. Roney, along with other members of the Game Protection Clubs of East Saginaw and Bay City, Michigan, ventured to Petoskey, Michigan, in April 1878. "Little did those clubs realize the task assumed," Roney wrote in the *Chicago Field*. Roney, William Fox, R. Fairchild and Del McLean already knew the names of some lawbreakers. One of those men was A. B. Turner, editor and owner of a Grand Rapids newspaper and president of the Kent County Game Protection Club.

Petoskey was filled with talk of pigeons and with pigeoners—many of whom had traveled from distant states. "The 'pigeoners' hurried hither and thither, comparing market reports, and soliciting the latest quotations on 'squabs,' " wrote Roney. "A score of hands in the packing-houses were kept busy from daylight until dark. Wagon load after wagon load of dead and live birds hauled up . . ." At a hotel in town, the citizen-enforcers met woodsman "Uncle Len" Jewell, who agreed to assist them in reconnoitering the huge nesting, which Roney estimated to cover about 100,000 acres.

Roney and his men left to find the nesting, more than 15 miles away, and when they arrived they "stood and gazed in bewilderment . . . On every hand the eye would meet these graceful creatures of the forest, which . . . darted hither and thither with the quickness of thought . . . In every direction, crossing and recrossing, the flying birds drew a net work before the dizzy eyes of the beholder, until he fain would close his eyes to shut out the bewildering scene." Then they heard the sound of falling trees. In another part of the forest, they saw nests flying everywhere as crews cut down trees and squabs tumbled helplessly to the ferns. Only about half the squabs were alive. The others had starved because so many adult birds had already been taken.

Fox, Fairchild and McLean headed back to Petoskey in a fruitless attempt to obtain arrest warrants for the clear violation of the ban on hunting within a nesting colony. Meanwhile, Roney and Uncle Len arrived at another hotel nearer the scene of slaughter. The professor hoped to learn of the location of illegal nets being operated by out-of-staters, but no one would talk. So, in the lobby, Roney "stretched himself on a bench, and was soon to all appearances in the embrace of gentle Morpheus, emitting snores that would have done credit to a full grown porpoise." Oblivious to the ruse, the pigeoners never suspected that Roney was taking in every word of their conversation.

Like the out-of-state hunters, Petoskey locals were equally uninterested in talking about pigeons. "At another house across the way, which harbored two or three trappers," according to Roney, "an old hag told us 'there wasn't no nesting anywhere around there that she knew of,' while she dressed and fried before our eyes some fresh caught birds, and the incessant twittering from the nesting not half a mile away was plainly audible." Roney soon learned it was more effective to pump children for information.

The next day Len and the professor found some illegally placed nets, watching in horror as hundreds of pigeons were caught and killed by "a stalwart pigeoner up to his knees in the mire and bespattered with mud and blood from head to foot." Roney observed that "when all were dead, the net was raised, [with] many still clinging to its meshes with beak and claws . . . and were shaken off." The pigeoner massacred 984 pigeons that day and did so "within 100 rods of the nests and in plain hearing of the nesting sounds, instead of two miles away, as the law prescribes."

This, then, was what Roney and his game-protection comrades wanted: more clear evidence of violations. "The next day [I] swore out a warrant and caused the arrest of the offender, who could not do otherwise than plead guilty, and had the satisfaction of seeing him pay over his fine of $50 for his poor knowledge of distances."

Though the professor's team had become the target of threats from locals and commercial pigeoners, Roney next set his sights on A. B.

Turner, the Kent County editor and "sportsman." In the Petoskey area, Turner was considered an out-of-county bigwig, and it helped that several people had independently established his guilt in shooting pigeons at the nesting and shooting at flocks, simply to count how many birds he and his party could kill. The local county prosecutor, however, refused to allow the sheriff to serve an arrest warrant. Roney, instead, had a civil summons issued and lined up willing witnesses, but the jury found Turner not guilty on a technicality, an outcome Turner had predicted. Meanwhile, poor Uncle Len, forgetting that it was not yet fishing season, was fined for catching a bull trout.

In pursuing their goals, Roney, Fox, Fairchild and McLean received the help of a railroad agent named M. F. Quaintance and two sheriffs, Tucker and Ingalls. Ingalls helped Roney's crew drive out hundreds of Native Americans encamped at the nesting, who also were killing squabs. (It is interesting that the commercial pigeoners apparently did not try to drive the Indians out; if the market-hunters had perceived that the pigeons were becoming a limited resource they likely would have opposed the Indians' presence.)

Perhaps the greatest accomplishment of the East Saginaw and Bay City Game Protection Clubs was that their men forced a steep decline in the market for Petoskey pigeons. The number of barrels shipped daily plummeted from 60 to 8. In his written account of the incident, Roney slyly does not explain how this happened. Apparently the professor and his cohorts spread the rumor that the Petoskey birds had eaten poisonous berries and, therefore, were not fit for human consumption. Newspapers across the country ran the story.

Roney estimated that *one billion birds* had been killed or wasted during the Petoskey nesting from March to August 1878. But pigeon-and-game-dealer E. T. Martin responded to Roney's charges in an article of his own in the *Chicago Field*, accusing Roney of, among other things, spreading the poison-berry story, of not caring for the poor (who benefitted from the pigeons, it is true) and of greatly exaggerating the number of birds shipped and slaughtered. Martin put those numbers at 1 million and 2.2 million respectively. (Biologists Blockstein and Tordoff calculate that perhaps as many as 10 million pigeons died at Petoskey in 1878.) Martin concluded in a huff, "The pigeon is migratory, it can care for itself."

The decades of the 1870s and 1880s, despite Martin's faith, saw a precipitous decline in the population. The numbers of nesting birds declined from millions to thousands, and most late nineteenth-century nestings ended in killing or abandonment. Pigeons became far more wary in these last breeding attempts, often departing immediately when humans began to shoot. Apparently, by 1886, there existed only two flocks of pigeons, one in Oklahoma and one in Pennsylvania. In the Northeast, the last large colonial nesting occurred in 1851 and the last nesting in 1880. In the mid-Atlantic states, the last large colonial nesting occurred in 1868 and the last nesting in 1889. For the Midwest, the years for each were 1855 and 1893. For the Great Lakes region, the dates were 1885 and 1894, according to ornithologist James Greenway.

All this elicited some public concern about the eventual extinction of the Passenger Pigeon, but the scarcity of the birds motivated hunters even more than had abundance. A story in the September 11, 1885, *Racine Times* about a recently arrived flock roused 500 men to the locale in just one hour.

By the 1890s, Americans counted pigeons by the thousands, by the hundreds, or by the tens, if they saw them at all. (Pigeons lived about 10 years in the wild, so the persistence of some birds long after breeding success had been overwhelmed by mortality is not surprising.) One Arkansas netter shipped a paltry 2,000 pigeons to Boston in 1891. In 1895, a St. Louis game dealer had lived through his second year of no pigeons. So by the time the Michigan legislature declared in 1897 that "Pigeons will not be shot by any body, anywhere, any time," the lawmakers had made the perfect law. Everyone could obey it because the pigeons in Michigan were virtually gone. (Incidentally, North American shorebirds filled the market gap left by the lack of pigeons. The Eskimo Curlew is thought to be

AMONG THE PIGEONS.

A Reply to Professor Roney's Account of the Michigan Nestings of 1878.

—BY—

E. T. MARTIN,

In the CHICAGO FIELD, Jan. 25, 1879.

E. T. Martin's Headquarters at Boyne Falls, Michigan, during the
Nesting of 1878.

*E. T. Martin's circular, replying to H. B. Roney's charges
against him and the Petoskey pigeoners.*

extinct because of commercial hunting, though the species might barely hold on today.)

Chicagoan Edward B. Clark hadn't seen a Passenger Pigeon since childhood, when, on a morning in April 1894, he stood in Lincoln Park, looking at a lone male pigeon resting in a maple tree. The bird faced east as the sun climbed higher. Clark recalled how "every feather shone, and the bird's neck was gem-like in its brilliancy . . . I watched the pigeon through a glass for fully ten minutes. A park loiterer approached and said he wished that he had a gun . . . That man had no soul above pigeon pie." Clark flushed the bird, hoping he would fly away from the dangers of civilization, but the Passenger Pigeon flew "straight toward the heart of the smoky city."

One day in 1905, C. F. Hodge, a Clark University professor spearheading a campaign to discover Passenger Pigeon nests, glanced up from his garden in Worcester, Massachusetts, and saw, he thought, about 30 Passenger Pigeons flying overhead. He waved his hat, shouting, "The Passenger Pigeons are not extinct!" But the leaflets he had helped distribute on how to identify the pigeon—and the rewards for finding live nesting birds—yielded nothing but mistaken reports for the American Ornithologists' Union. The group officially ended its search on October 31, 1912.

People kept looking, of course, and some claimed to have seen Passenger Pigeons, though they probably were looking at Mourning Doves. The Band-tailed Pigeon, a bird of the American West, even fooled old-time pigeoners. One former hunter claimed to have seen a Passenger Pigeon . . . near Los Angeles! Sightings of "Passenger Pigeons" thus continued well into the twentieth century. It's possible that some were authentic—perhaps they were released or escaped stool pigeons—but none of these reports could be authenticated.

One especially intriguing sighting appears to have been overlooked. A. W. Schorger spent years researching late records and claims for pigeon sightings, but did not include in his book Edward A. Preble's account of a reported flock of Passenger Pigeons in West Virginia in 1907. In early December 1907, on behalf of the U.S. Biological Survey, Preble found himself investigating rumors and garbled newspaper accounts of a rare flock of pigeons. He traced the source of the allegations down to a man named A. B. Elbon of Webster Springs, West Virginia. Preble, in his papers at the Smithsonian, reported that on October 26, 1907, Elbon and "a companion were hunting on Elk Mountain, near Webster Springs, [when] their attention was attracted by a loud rushing sound which at first they were unable to locate." Preble continued:

> In a few seconds, however, a flock of wild pigeons flew over their heads at a height of about 200 feet. The birds were in a loose scattered flock, not "lined up" . . . but flying as if looking for a suitable feeding ground. They were going nearly due west. [Elbon] estimated the number at the time at about 500. He did not shoot at them. In a few seconds the birds disappeared from sight.

Elbon had seen migrations of Passenger Pigeons in years past. He knew the birds. And, though no one else saw this flock, Preble said he believed Elbon, a longtime sportsman. Could Elbon have been mistaken? He was "a man perhaps 50 years of age," and so may have had less-than-perfect vision. Had he seen instead a large flock of Mourning Doves?

For many years, people could not believe the pigeon had vanished. Perhaps the birds had migrated to Australia or Bolivia. Others thought the pigeons had died from a virulent disease. Even today, people wonder if perhaps the birds had all plummeted into the Gulf of Mexico. That drownings of pigeons had occurred elsewhere made this supposition seem plausible, and there were reports of a mass drowning in the Gulf late in the species's life. But the reports were not credible. In 1892, an Osage guide named John Aurochs gave his reason for the disappearance to his camping companion, John French, a man who was obsessed with the life and loss of the pigeons. The two were camping in Oklahoma, where

French had come to scout out timber for making gunstocks in England. Beside the fire, Aurochs quietly confided to French what had happened to the pigeons: The birds had just given up, the pigeons had "abdicated" their reign of the sky to the Northern Cardinal.

In actuality, human activity—habitat destruction and hunting—ultimately caused this extinction. Wide-scale clearing of forests had reduced the availability of mast and thus limited options for the huge breeding colonies that had come to depend on that food source. Though perhaps enough mast was still available to support smaller nesting groups, feeding efficiency declined because the birds could not locate mast as easily as before. Biologist David Kirk notes that beech trees, the pigeons' favorite source of nuts, occupied lowlands—the areas that settlers first cleared for farms. And upland mast-bearing trees, such as oak and chestnut, provided wood fuel. "The clearance of forests by Europeans reached a peak in 1880," he says, "and this coincided almost exactly with the sudden increased rate of decline in the Passenger Pigeon."

Cutting forests also had two other effects on the species: First, it meant that brushy plants took over the ground in second-growth forests, making it more difficult for pigeons to forage; second, Kirk points out, beech trees yield mast for the first time when they are 40 years old, but most trees at least that old had been cut down during the pigeons' final decades. Younger beech trees couldn't produce mast for many years to come. Hunting, as well, disrupted roosts and nesting colonies. Recruitment of juveniles into the population plummeted because young were directly killed and because breeding adults were destroyed. Without the huge numbers of birds in the colonies to stimulate breeding—and offset nonhuman predators with sheer numbers—the pigeons became reproductively dysfunctional and isolated. The remaining, fragmented populations, in dwindling groups and pairs, became prone to accidents of predation, bad weather and inbreeding. (Stray Passenger Pigeons began to associate with Mourning Doves, but the possibility that Passenger Pigeons and Mourning Doves bred in the wild is virtually nil.)

What, in the wider culture, is the legacy of the Passenger Pigeon today? An insult, "stool pigeon" (the origin of which most people do not know), and a mistaken term, "carrier pigeon" (an error which many make when they mean to talk of Passenger Pigeons). A handful of monuments do commemorate the species, including a bronze plaque set in stone on Sentinel Ridge in Wyalusing State Park in Wisconsin, overlooking the junction of the Wisconsin and Mississippi rivers. This marker inspired Aldo Leopold's moving essay, "On a Monument to the Pigeon." One mile west of Oden, Michigan, on U.S. 31 in Littlefield Township, stands a sign describing Passenger Pigeons and their fateful nesting near Petoskey in 1878.

We are too far removed from the pigeon era for anyone living now to remember the species directly. But memories of *the people* who knew pigeons do persist. When I was interviewing the painter Don Eckelberry about his experiences watching the lone Ivory-billed Woodpecker in the Singer Tract, he told me this story. Decades ago, when Eckelberry worked as a naturalist in a forested park outside Cleveland, he met a retired farmer, a man in his eighties or nineties. Eckelberry remembers that the farmer talked "about how he used to see the blue pigeons. 'They weren't the little mourner doves,' he'd say. 'They were big blue pigeons and came in great bunches. They'd land at the edge of the forest where the beech mast was and would keep moving ahead, the back of the flock flying up and over the rest of the flock. Sort of rolling ahead, across.' He didn't say if he hunted them. He didn't say anything about eating them. But I have no doubt about his authenticity."

I never expected to speak with someone acquainted with a man who had *heard*, who had *seen* Passenger Pigeons. There is a distance in this telling, of course, but the story makes me feel, sometimes, as if I stand in a part of that dark beneath their wings, the dark of great, mislaid passages.

The Boy and the Pigeon

What a strange thing is memory, and Hope. One looks backward, the other forward.

—Press Clay Southworth, journal entry, 1975

BESIDE A GLASS DISPLAY CASE, IN A DESERTED HALL OF STUFFED ANIMALS, next to an emergency exit, I read this sign:

On March 12, 1900, 14-year-old Press Clay Southworth shot an unusual-looking bird on his family farm near Sargents in Pike County, Ohio. Recognizing the bird as a passenger pigeon, his parents had Press take it to an amateur taxidermist. She mounted the pigeon, using shoe buttons instead of glass eyes. Eventually authorities determined that this was the last documented wild passenger pigeon.

Press Clay Southworth, I whispered. What a lovely name. I read the sign again. Who was he, I wondered, and how could I learn of his life? I had come to Columbus to the Ohio Historical Center's museum to see this bird, this specimen of the last free-roaming Passenger Pigeon and to learn more of its death. Now I had the name of the shooter and a few other particulars of the incident.

Then I realized: The date on the sign, March 12, 1900, was wrong. The authoritative sources about this shooting all claimed it took place on March 24, 1900. So with the *incorrect* date emblazoned on this plaque in front of me, I wondered if I could trust what *should* be definitive, this label text in an impressively curated and expansive historical museum. I took notes and scribbled questions. Then I set aside my doubts to give full attention to the pigeon nicknamed "Buttons."

The first thing I thought when I saw her was, *She's not a very pretty pigeon.* Buttons looked dingy. On the right side of her neck a trace of purple iridescence glistened, but dust had settled on her black shoe-button eyes. Her body mingled browns with moments of white. I peered at her pinkish legs and a beak and claws the color of dark dress shoes. She was displayed in a somewhat unrealistic posture, front-heavy and horizontal, which accentuated her long body and tail. A curl of dust perched on her back, and a barb of tail feather stuck up like a sickle. A hole showed at the base of the white patch just below her beak. Was that where she had been shot? A metal wire protruded just beside the left leg.

Buttons was kept company by two Ivory-billed Woodpeckers, two Carolina Parakeets, some bivalve mollusks, a handsome male Passenger Pigeon, a trapper's net and a stool-pigeon perch. That leather circular perch looked to me not unlike a Plains Indian dreamcatcher. Lights illuminated Buttons's back and her right side, while the rest of her stayed in permanent shadow.

For decades after Buttons's shooting, the full story of the incident was also shadowed. For years, anyone interested in the event—and very few people were, despite its historical importance—had to rely on fragments of details or even sheer conjecture.

I first encountered the fact of this shooting when I read a brief sentence about the event in the *Check-list of North American Birds*, which accepts, in all its recent editions, the March 24, 1900, date. Yet even this source contains an error; the *Check-list* has a typo, misnaming the site of the shooting as Sargento instead of Sargents.

Initially, the incident was summarized by a German-speaking pastor and naturalist named W. F. Henninger, who, in 1902, wrote an article on south-central Ohio birds for *The Wilson Bulletin*. Though I could never locate Henninger's papers, which might have yielded more information, it seems clear that he personally inquired into the shooting. In his article, Henninger named the amateur taxidermist who had prepared Buttons—a Mrs. Barnes, wife of Pike County's ex-sheriff—and he said that a "small" boy had shot the bird near the Scioto and Pike County line on March 24, 1900. Given Henninger's proximity to the event, the details he shared are considered definitive. Reverend Henninger thought this shooting, supported by the preserved specimen, was "the only authentic record [of a Passenger Pigeon] for twenty years."

The venerable A. W. Schorger agreed with this conclusion. Based on Henninger's authority, Schorger dismissed the claim of a writer named Henry Bannon who reported that the bird was a *male* shot in 1899, a mile and a half east of Wakefield, a small town near Sargents. The preserved bird was clearly a *female*, not a male, as Bannon had asserted, and no one at Ohio State University, where the specimen was kept, accepted Bannon's baseless claim.

As for scholarly sources, Henninger's article and one paragraph in Schorger's book were it—remarkably little given how much had otherwise been written about Passenger Pigeons. The world's last-known wild Passenger Pigeon seemed to merit only a passing mention. And though Henninger "had ample opportunity to investigate the accuracy of the [shooting's] circumstances," wrote Schorger in 1955, those very circumstances, the specific textures of the incident and that day were never retrieved.

At least until 1965. That's when Allan W. Eckert published a novel about the extinction of the species called *The Silent Sky*. Eckert imagined a male pigeon landing in "a cluster of oaks and hickories along the east bank of the Scioto [River] . . ." After unsuccessfully trying to bag a chipmunk, a boy watched the pigeon walk across leaf litter. Finally it moved within range, and the boy lifted his BB gun—a Christmas gift, according to Eckert—and shot the pigeon in the head. The boy felt a "pride . . . boundless," when he presented the corpse to Mrs. Barnes. A *Time* magazine review of the novel repeated these particulars, which soon were taken as genuine in various popular magazine articles.

Since the fact that Buttons is a female specimen was easily verifiable even in 1965, I doubted that I could trust other details that Eckert offered, especially his assertion that the shooting took place along the Scioto River. But, for a long time, Eckert's fictive narration was the only account I knew of that presented specific circumstances of the incident.

Then I saw an obscure newspaper column that added several significant facts to the story of Buttons. On March 2, 1970, *Columbus Citizen-Journal* columnist Ben Hayes wrote that "Press Southworth, 84, retired locomotive engineer of 92 Powhatan [Avenue], has declared that he shot the bird 70 years ago this month—dispelling a lingering mystery about the gunner . . ." Southworth used a 12-gauge shotgun, not a BB gun, to shoot the Passenger Pigeon from a tree on the family farm, not by a river. The boy had known Mrs. Barnes from at least one previous encounter: "[a] few months before [Press Clay] had bagged a white crow which Mrs. Barnes stuffed for him," wrote Hayes. This column, I learned, became the basis for the sign at the Ohio Historical Center naming Southworth, even though, somehow, the wrong date appeared on the museum's placard.

Eventually, I found that the museum still had the accession book that recorded Buttons's arrival. Someone had noted the bird's acquisition into the collection with the following entry: "1915, Feb. 27. Mr. Clay Barnes, of Waverly, Pike Co. presented mounted specimen of Passenger Pigeon. This is the one with shoe buttons for eyes. Killed March 22, 1900 at a siding at Piketon, O by a boy named Southworth. Mounted by Mrs. Barnes." This information gives us yet another incorrect date and a vague location reference, even as it names—and confirms—Southworth as the shooter. Staff at the Ohio Historical Center seem not to have noticed or followed up on the naming of Southworth in this note. Perhaps the book had been forgotten or misplaced for a time. Schorger apparently did not know of the volume; had he known, I suspect he would have contacted South-worth and interviewed him.

The idea of donating Buttons to the museum was first broached in an October, 1914, letter from H. C. (Clay) Barnes, husband of the taxider-mist, to Professor William Mills, the distinguished curator of the Ohio State Archaeological and Historical Society:

> Dear Sir, I can not recall your initials is the reason I addressed you as above. I have a mounted Carrier Pigeon [*sic*]. I can not take care of it and it will be lost, if I keep it. Has the University a collection of birds[?] If it has not please advise me what to do with it. I will send it up if it has.

Mills soon replied, "I wish to say that if you have a mounted Carrier Pi-geon, we will be glad to have it." Mills also noted that the society did not have a natural history collection as such, but would soon, "as we are get-ting so many natural history things . . ."

After Barnes shipped Buttons to Ohio State University the museum staff put the bird on display. Later, Buttons was on loan to the Cincinnati Zoo, which, in its own sign, claimed that the bird had been shot by a "grainery." (Just how that detail originated is a small mystery.) In 1992, Buttons went back on display in the renovated Ohio Historical Center in Columbus.

As I stood that afternoon next to Buttons, separated from her by a plate of glass, I knew some of this. I knew that what one adolescent boy had done constituted the final skirmish in our ancestors' campaign against the pigeons; most farm boys and ornithologists would have behaved no dif-ferently. Yet, even after learning more about the boy and the pigeon—reading that 1970 newspaper column in particular—I still felt the itch of incompletion. No one had the full story, if it existed.

"Buttons," the world's last known wild Passenger Pigeon, when she was on display at the Cincinnati Zoo. The date is unknown.

Our memory of Passenger Pigeons, now mediated through words, and our memory of Buttons and of this vague Ohio boy . . . all demanded that I try to discover more, that I work to discern whatever else I could. Though I now had Press Clay Southworth's name, I needed to amplify his existence. Who was this boy? Who remembered him? Though I had seen Buttons, I needed to see the land on which she had died. Where was that farm?

I needed to explore, to scout the territory of Pike County, and find ways to answer these questions or, at least, determine if the questions were, finally, unanswerable.

So in the summer of 1993 I went to Pike County. My wife and I drove on backroads into Appalachian foothill country—a section of Ohio that had not been glaciated—past beaten-down houses and trailers, fields of crops and fair stretches of woods. The 444-square-mile county was organized in 1815, some 20 years after the first pioneers arrived in the area, and is now home to about 24,000 people. We arrived at the county seat of Waverly, about a 90-minute drive south of Columbus. An old Ohio–Erie Canal town founded in 1829, Waverly is charming, if faded. Twentieth-century billboards and franchise restaurants corrode the town's red-brick Victorian ambience.

Waverly's mayor is a gray-haired and friendly man named Blaine Beekman. Despite his prodigious knowledge of the area's past (Beekman is a local historian and part-time lecturer at Shawnee State University), he didn't know the story of Press Clay Southworth and the last wild Passenger Pigeon. Beekman could tell me of the Barnes family, however, and said that, in 1900, they lived in Sargents. Had the Southworths been renters on the Barnes property? he wondered. I wrote down the names of Southworths still living in the area, but decided I wouldn't contact anyone until I had done as much homework as possible.

Rushed for time—Elizabeth and I were heading to Maine for a conference—I took ample notes on Beekman's reasonable suppositions about the location of the Southworth land in 1900. And I got a good feel for the county's topography and landmarks, such as the U.S. Department of Energy uranium enrichment plant (which produces reactor fuel). The plant is located across from a sand-and-gravel pit, part of what little remains of Sargents. If the Southworths had lived on property now controlled by the D.O.E., I felt certain I would not get permission to visit.

At the volunteer fire department in nearby Wakefield, just two miles south of Sargents on U.S. Highway 23, Beekman and a man named Teddy West pulled out a 1952 property map. They pointed to parcels of land owned by a W. G. Southworth, not far from where the old Barnes house had stood. There had been Southworth land on the east side of the old Columbus-Portsmouth Road, now called Wakefield Mound Road, and just north of Salt Creek. Beekman now felt sure that the Southworths had been hired hands or tenant farmers somewhere near Wakefield.

Theories and leads, but nothing definite. At the local library, I consulted the 1900 census, taken in June of that year, and excitedly found an entry for 42-year-old blacksmith Orlando P. Southworth and his 38-year-old wife, Maria M. Orlando, renting land on farm 113 in Scioto Township, according to something called the "farm schedule." Living with Orlando and Maria were their children George J., 21; Calvin W., 19; Gracie B., 16; Pressley C., 14; Arley W., 10; William H., 7; and Opal M., 2. *Pressley C., 14.* That was him, Press Clay Southworth, one of Orlando and Maria's sons.

In the library, I could touch history—part of the Waverly canal stonework forms a basement wall for the building—and I felt hopeful that I might find more than just the census record. I searched next for local newspaper accounts of the shooting of a rare wild Passenger Pigeon, but found none. Because these birds in their heyday were creatures whose movements were reported by newspapers across the eastern and midwestern United States, I found it strange that the local *Waverly Courier-Watchman* published no notice of the occurrence of a straggler pigeon. Surely that would have been news. After all, most adult residents of the area would have seen or have known of the great flocks from years before. Pike County was in the old heart of pigeon country. In the Wednesday, March 28, 1900, *Courier-Watchman*, four days after the shooting, the editor remarked upon a visit by ex-sheriff Barnes, husband of the amateur taxidermist, but said nothing of Buttons's death.

Both Barnes and the editor of the *Courier-Watchman*'s competition, the *Waverly News*, were Republicans and, therefore, probably on good terms. Like most newspapers of the day, the *Waverly News* published all kinds of talk. If Barnes's visit had occurred after his wife had received the dead pigeon, and if Barnes thought it worthwhile to mention, surely the editor would have written it up, even in a sentence or two. (It's not entirely clear when Barnes was visiting Waverly proper; the March 22 issue of the *Waverly News* has the ex-sheriff conducting visits in the prior week, so his courtesy calls may have happened before the incident.) But perhaps I missed an article in one of the March, April or May issues. Perhaps an editor cut notice of the shooting for lack of space.

While I scanned issues of the *News* and the *Courier-Watchman*—finding an item about a flock of Canada Geese being confused by electric lights in town, a remark upon the arrival of Purple Martins, a comment about fish biting in the canal and a report on Miss Nellie Higgens's piano recital—my wife continued to look at the 1900 census. There Elizabeth found something that for many months I failed to pay attention to.

All that time I had been trying to find copies of the "farm schedule" mentioned in the 1900 census. I knew that because Orlando was not a property owner, I could not rely on finding his tax records to tell me exactly where he lived. Such records might not even exist. Orlando was renting farm 113, the census told me. So, I reasoned, if I could get a copy of the farm schedule, I might find a description of farm 113's location. I contacted historians, county, state and federal offices, but never found such a document. Meanwhile, in Elizabeth's notes was the answer all along: Another Southworth, the elderly Calvin, also of Scioto Township, *owned* farm 113, free and clear of mortgage. *Orlando was renting land from his father, Calvin.* Because the census did not list Orlando as living in the same household with his father, Orlando and his family had to have been in a separate dwelling somewhere on Calvin's land.

By the time I realized this, I was back in Kansas. And the Southworths I contacted who had signed the library genealogy register could offer nothing about the life of Press Clay. So, by phone, I bothered a slew of Pike County offices to see if property and tax records from 1900 might be available for loan or copying. They were not. Voter registration records? Not available. School enrollment records? Not available.

Until I could return to Pike County to unearth property tax records for Press's grandfather Calvin, I was stymied in my search to locate the farm on which Buttons had died. The prospect of finding those records seemed dubious and, in more than one way, distant. "Oh," a secretary told me, "I doubt we still have records that go back that far. If we do, they'd be in the Vallery Building basement." She paused. "And it's a real mess down there."

By 1920 there occurred no record of Press Clay in Scioto Township, and farm 113—wherever it was—had become the residence of a Charles Brigner and his wife, Odessa.

A scrap of paper provided the lead I needed. The curators at the Ohio Historical Center had found in their files a photocopy of a business card for a realtor named Ted Kruse, who, they said, was Press Clay's grandson. Kruse had left his business card with a photocopy of the 1970 Ben Hayes column about the Buttons shooting. This, I thought, might be the breakthrough. Perhaps Ted Kruse could spare me a tedious deed search for Calvin Southworth's land. Surely he could tell me a little about his grandfather, Press Clay.

But the photocopied business card was out of date by some 20 years or more. Realtors contacted in Grove City, Ohio, near Columbus, could not tell me where Ted Kruse lived. Through a realty association, I found that he had worked with another company, but not since 1993. Phone numbers weren't operative. The post office returned my letters. Exasperated, I located everyone in the Columbus area whose name was similar to Ted Kruse's and sent off letters explaining my research. And I waited.

Then one day, in fall 1998, I came home to a message on my answering machine: "Hello, Christopher, this is Mary Kruse, Press Clay Southworth's daughter." She explained that one of her grandsons had received my letter and passed it on. I called and we arranged a time for me to visit. I hung up, silently thanking Ted Kruse for leaving his business card at the Ohio Historical Center years ago. Had he not, it's likely that the full story of March 24, 1900, would have been forgotten.

Press Clay Southworth, in 1904, four years after he shot Buttons.
He is 17 or 18 in this photograph.

On a cloudy, warm and humid afternoon in late November 1998, I sit in the living room of Mary Kruse's modest home on Brinker Avenue in Columbus. On the wall hangs a photo of Press Clay Southworth from 1904, when he was 17 or 18.

Mary, Terry, Ted and his wife, Vi, tell me wonderful yarns, facts and anecdotes in no particular order. I learn that Press Clay worked as an engineer on the Chesapeake and Ohio Railway for years, starting his career in 1905 on the Hocking Valley run and eventually becoming an Assistant

Grand Chief for the Brotherhood of Locomotive Engineers. He even served on the National Railroad Adjustment Board, the union's grievance office; for a time, he lived both in Chicago, where the board was located, and in Columbus, where he spent most of his life. For a man with a sixth-grade education, Press Clay went far.

The Kruses tell me that Press Clay's brother Billy was a famed major-league outfielder and manager, and that Press Clay had played some ball himself. When Casey Stengel didn't show for an exhibition game in 1919, the stocky railroad engineer subbed for the legend, borrowing shoes from Billy and donning Stengel's Pittsburgh Pirates uniform. Press told a sportswriter that the team provided him with a contract written on toilet paper. He had two hits for the day. For the Columbus Dry Goods squad, Press once struck out 17 men.

Despite this prowess, he stayed with trains. "Orlando [Press's father] didn't think there was any livelihood in ballplaying, no future for it," Mary says. "The future was in the railroad!"

Twice married, Press outlived both wives, grieving deeply for each, especially his second wife, Mary, to whom he had been a husband for 63 years. He outlived two of his children and all of his brothers and sisters. He died in 1979 at age 94 after what was, it appears, a full and good life. Press was a strong and polite man, witty and usually patient, though Ted recalls more than one sharp comment directed at him when he was helping his grandfather paint the house and kept missing spots.

According to the 1900 census, Press Clay Southworth was born in Ohio in August 1885, but according to Press himself, he'd been born on August 27, 1885, in Clay County, Nebraska. "That's where his middle name came from," Mary says, holding her hands on her lap. "It's all in the blue book."

"The blue book?" I ask.

She gives me a copy of a blue-covered scrapbook full of stories, newspaper clippings, pictures and poems of Press Clay's, all compiled by his granddaughter Peggy Jean.

Press Clay Southworth at age 65.

In the scrapbook, Press wrote of his boyhood in Nebraska, on 640 acres of homestead land that Orlando staked claim to after moving from Ohio with his wife, Maria. "Coyotes and antelope were plentiful," Press recalled. He remembered attending a one-room schoolhouse in Clay County, southeast of present-day Grand Island, beneath a wide prairie sky. Though he doesn't write of them, he would have witnessed the migration of Sandhill Cranes each spring, when wheeling flocks of these large, elegant birds fly high above the earth and rest at night in the shallow waters of the Platte River.

Press remembered how herds of cattle grazed on his father's ranch not far from the Little Blue River. How box elder trees lined the watercourse.

How the pools created by beaver "afforded good fishing." He reminisced about the chirpy calls of prairie dogs.

And he couldn't forget the Indians who arrived from South Dakota to camp on Orlando's land in 1893:

> They were equipped with tents and ponies, bringing their squaws and pappooses [sic] with them. They had many dogs that they kept well fed for food, cooking them for food, in large kettels [sic]. My Father took we three older boys to the Indian Camp. The Chief took us into a tent where squaws were cooking a dog. My Father told them they would have to leave. However they stayed.

This laconic retelling doesn't fully capture what must have been a tense confrontation. The tribe—Blackfeet, according to Ted and Terry—slaughtered two of Orlando's steers before heading back to the reservation farther west. The Indians also raided the Southworth cabin.

While in Nebraska, the family befriended Buffalo Bill Cody, the famed U.S. Army scout and hunter for Union Pacific railroad crews. Cody participated in the great bison slaughter in the American West that pushed the species to near-extinction. He also formed Buffalo Bill's Wild West Show in 1883. "My mother often prepared a nice meal for [Cody and his men]," Press wrote at age 93, a year before he died. "Cody always called [my father] Lan for short."

Unfortunately, droughts set the Southworths back, and a fire destroyed their ranch home. Press, who wore a sister's gown as a young boy and therefore fussed about having to don pants, unhappily clambered into a covered wagon one day, in whatever it was his parents told him to wear. The Southworths were moving back to Ohio to start over again.

In Ohio, their old friend Buffalo Bill brought his show to Portsmouth. "My father and I visited him in his tent in 1898," Press wrote. "They had a long visit going over old times and a good handshake when we left. He ruffeled [*sic*] up my hair and said always be a good boy." Press would say, years later, to his grandchildren, "Shake the hand that shook the hand of Buffalo Bill . . ."

I flip through the blue book and listen as Mary, Terry, Ted and Vi banter easily as they tell these stories. Mary is a short, gracious and unpretentious woman, given to frequent laughter. She sprinkles her sentences with interjections of "yes, right" and "well," delighted to be talking so much of her father. She wears a full head of wavy, pepper-and-silver hair and is, in a word, spunky. Terry, who looks not unlike the actor Robert Duvall, and Ted, who reminds Vi of the 1970s singer Steve Lawrence, rib their good-natured mother throughout the conversation. When I ask a question that Mary has trouble answering, when she clasps her hands together and exclaims, "Oh golly, I wish I could remember more about that," her sons quit joshing. "You're doing a good job," Terry says quietly.

I ask Mary about Press Clay's recollections of Pike County, but there is little to tell. Press didn't talk as much of that place, she says, as he did of Nebraska and of Washington Courthouse, another Ohio community in which Orlando's family lived. Though I regret not knowing more about her father's days in Pike County, I am glad to learn that, as Terry observes, Mary Kruse holds her hands the way her father did. She has his quick step, too. Such characteristics are themselves a form of remembrance.

More than a half-century after he shot the pigeon in Pike County, Press Clay Southworth found the bird at the Ohio State University Museum. One summer day in the early 1950s the family visited the museum on High Street, having heard that a pigeon with button-eyes was on display. Her father hadn't known it was there. "This was where we first saw his pigeon," Mary says.

Terry relates the story of how his grandfather twice took him pheasant hunting in 1959. "I'm sixteen and he's seventy-four," Terry says of that year. (His grandfather would keep hunting for another 10 years.) Terry continues:

Grandpa wanted to do something special with my first two [pheasants] so he skinned them out and we ate the meat. After the hides dried, grandpa placed them on a finished board with their wings in full flight . . . [and] put white buttons in the pheasants' eye sockets and sewed them in through the skull. Next he took black paint and painted black dots for eye color on the buttons to make them look like the birds had [their] eyes wide open . . .

For most of his life, Press Clay treated the story of Buttons like this, as a family matter, something interesting to tell and acknowledge only to those nearest him.

Perhaps he knew from whatever sign hung at OSU in 1959 that Buttons was the last wild Passenger Pigeon in the world. Nonetheless, another nine years would pass before the full significance of the shooting seemed to come clear to Press Clay. In his early 80s, he looked back on his life and understood just what an important role in history he had inadvertently played.

Mary Kruse remembers that in 1968 an article on Passenger Pigeons appeared in the February-March issue of *Modern Maturity* magazine, an article that repeated the story of a young Ohio boy who had shot a Passenger Pigeon with "a Christmas BB gun . . . in an oak grove." The writer noted that this bird was the last wild pigeon. That was when, it appears, Press Clay realized what this shooting meant. Soon after seeing the article, Press sent the magazine a letter describing the event, though the editors didn't publish it in the issues I checked. Later, he contacted the Columbus newspaper writer Ben Hayes, though Hayes did not quote from Press's letter.

Writing in longhand on lined paper, Press quoted the *Modern Maturity* article's paragraphs about the "little boy" with a gun. Then he titled his letter *The little boy now 82 yrs old tells his story*. This is what, in 1968, Press Clay Southworth remembered of March 24, 1900:

I lived on a hill farm in Pike County with my parents and brothers and sisters. On the date in question I was feeding cattle in the barn yard when I saw a strange bird feeding on loose grains of corn near the cattle and as I approached it flew into a large tree near by. I had been raised on a farm and was quite familiar with the various species of wild birds. However this bird was larger than a [Mourning] dove and its flight was quite different than a dove or pigeon [Rock Dove]. I hurried to the house, told my Mother about this strange bird. I was only 14 yrs old and not allowed to use a shot gun[—] only by permission. After considerable persuasion she gave me our 12 guage [sic] shot gun with one shell. I was pretty handy with a gun as I had hunted with my older brothers. I found the bird perched high in the tree and brought it down without much damage to its appearance. When I took it to the house Mother exclaimed—"It's a passenger pigeon." She had seen thousands of them when she was a small girl. My Father also identified it as a passenger pigeon and told me to take it to Mrs. Charley Barnes, wife of the former sheriff of Pike County, so she could mount it. Mr & Mrs Barnes lived on what was known as the "Old Scioto Trail" midway between Wakefield and Piketon, Ohio. I took the passenger pigeon to Mrs. Barnes who was delighted to receive it and she took me in her home to show me some of the many specimens she had mounted. This trophy is still well preserved in [the] museum at Ohio State University, Columbus Ohio.

Press C. Southworth, Sr.

92 Powhatan Ave

Columbus, Ohio

To Press Clay Southworth's account, I can add little, except to say that the weather was cool. The low was 39 degrees, the high 55. Press must have had a pleasant walk or ride from his home to Mrs. Barnes's house on the Old Scioto Trail—the former "highway" for the Shawnee Indians between Kentucky and Lake Erie, a path that traders and soldiers later used and that is now covered by U.S. Highway 23.

When, in the early nineteenth century, pioneer Virginians arrived in what was to become Scioto Township, they found a heavily forested land, cut through by Salt and Big Run creeks, with rich bottomlands prone to flooding and the hilly, thin-soiled uplands suitable enough for raising crops. The settlers also found Passenger Pigeons, feared them, ate them and told stories about them, one generation to the next. Just three months into a new century, Press may have thought of those days as he carried his latest treasure to Mrs. Barnes.

In 1900, when Orlando was renting land from his father Calvin and Press Clay was helping with family chores, most Pike County residents made their livelihood from farming and lumbering. Though manufacturing industries were developing, Ohio had nearly 280,000 farms raising rye, corn, wheat and other crops. Until a crash in the farm economy in 1920, Ohio agriculture was a prosperous enterprise as land values and commodity prices increased. Even with the cost of investing in new, labor-saving equipment, such as tractors, farmers could make good money.

Ninety-five percent of Ohio had been forested before Europeans arrived to clear out farms and harvest lumber. At the turn of the century, however, only about 10 percent of the state remained forested and such habitat consisted mostly of scattered, fragmented woodlots. In 1900, the Scioto River flowing south through Pike County carried with it the trash and waste of the 125,000 residents of upstream Columbus. Throughout the state, deer and turkeys remained scarce. Cougar, bear, wolf, elk, beaver, pigeon, parakeet, Ivory-bill, bison—all gone.

1900 was the year of the Boxer Rebellion in China, the killer hurricane in Galveston and American child laborers on strike in coal mines and factories. The year of the Paris Exposition was also the year that W. E. B. Du Bois saw the body parts of a lynched black man exhibited in Atlanta. Electricity, moving pictures and concrete roads promised progress and comfort. People grew increasingly unhappy with the war in the Philippines, which did not suit an increasingly prosperous and optimistic middle class more intent on leisure than conflict.

In parlors everywhere, people played the Maple Leaf Rag. And train travel continued to make vacations, including scenic nature trips into the West, more affordable and popular. President William McKinley, a native Ohioan, had just a few months to live before his assassination. Teddy Roosevelt would become the next president and would institute policies to better conserve wildlife, forests and other scenic wonders. Americans still talked of and remembered the great pigeon flights of the nineteenth century, not knowing that 1900 marked the end of the pigeons' wild existence.

I stand in the rank basement of the Vallery Building in Waverly, Ohio, and look disbelievingly at the scene Ted Wheeler shows me. Wheeler, the Pike County Auditor, is embarrassed by the chaos, how volumes and volumes of county records lie scattered pell-mell on shelves, stacked haphazardly on the floor and jumbled, as if tossed, atop a long, worn workbench. I look at boxes overflowing with yellow paper. Old furniture crammed against the dank stone walls. Twenty years ago temporary workers moved these records from the courthouse to this building and, well, Ted explains, the men didn't do a very good job. Indeed. I look back at the darkened section of the basement, then scan the lit portion of the room, this depository in disarray, this dump of facts from a previous century.

After Wheeler leaves, I scan the faded letters of dark brown, leather-bound volumes: *Township, Personal Property, Pike County, Tax. Assessment of Real Property. Auditor's Duplicate.* Year after year of books bigger than volumes of the *Oxford English Dictionary.* 1898. 1901. 1897. On and on with only brief interludes of chronological order. I step into a cramped aisle, feel my eyes and throat begin to itch, then lift up a volume of nineteenth-century county court records. On the top of the book's unopened pages, mold has grown a half-inch high. I open the book, paper crumbles, and I try to keep from destroying the pages while carefully turning them. A chaos of local history's slow dissolution spreads before me. Here, I'm sure, is my last chance to learn where Press Clay South-worth shot Buttons.

I close the court records, open my notebook and look for turn-of-the-century tax and property registers. As I am alone—who, after all, would want to be in the musty, calamitous basement of the Vallery Building?—I say aloud, sarcastically, "Good luck." Pushing aside a stack of dusty volumes, I clear the space I need.

Each book I pull from a shelf or lift from the floor unleashes a small cloud of dust accumulated over the useless years. I sneeze. I keep wishing the deed search that I had attempted earlier in the day had yielded more complete and more understandable records. I still don't know where Calvin and Orlando lived.

In *Township, Personal Property, Pike County, Tax* for 1900 I find records for the two men, but no property descriptions. I do learn that Orlando owned a dog. In the *Assessment of Real Property* in Scioto Township for 1900 I see an entry for "Barnes, H.C. & C.S." that describes the locations of their landholdings. But I find *no* Southworths in Scioto Township listed as owning "real property." Perhaps I missed an entry, but I double- and triple-check, dismayed. After all, Calvin owned his land according to the census. How can this be? Why is he not listed in this book?

Finally, I turn to the *Auditor's Duplicate.* I look everywhere, but the 1900 volume is missing. In the 1899 edition, however, I discover Calvin Southworth listed with four parcels in the township—and the locations *are* described. With help, I should be able to match these descriptions to my passel of current maps. Calvin owned nearly 190 acres total in 1899. But what if his holdings had changed? The 1900 records are gone.

Worried, I pull down the 1901 volume, but the parcels remained the same. So Calvin and Orlando must have been on the same acreage in 1900. *Now I can find the land.* I can narrow the scene of Buttons's shooting from an entire township to a matter of a few dozen acres.

It's nearly 4:30 P.M., and if I'm to find this land before sunset I need to see "the map ladies," as they call themselves. I bound upstairs and wrestle the huge *Auditor's Duplicate* volumes over a photocopier, then return the books to the basement.

At the county highway department Barb Barker and Lori Burkitt decipher the detailed property descriptions and correlate them to a current Pike County road map and a property map from about 1910. Barb looks up the current owners, whose names I write on the 1910 map. We agree that Calvin's quarter-acre lot and one-eighth-acre lot in Section 12 of Scioto Township were obviously too small to have contained farmland. Perhaps one of these small lots was home to a blacksmithing shop for Orlando. These two lots were separate from two large, contiguous parcels of 80 and 106 acres in Sections 30 and 31 of Scioto Township. Somewhere on those 186 acres, Orlando Southworth farmed. Somewhere in that area, now shaded in pencil on my maps, stood Orlando's house, his barn and barnyard, where Buttons died among Mourning Doves and cattle. And, Barb tells me, even though the property is some distance from what used to be the community of Sargents, the whole area would have been called by that name back in 1900.

This property is now a private inholding just within a small section of the Wayne National Forest. The Southworth acreage from 1900 sits more or less at the center of a rugged, hilly area bounded by Salt Creek Road, Rapp Hollow Road and Big Run Road.

After stopping at the library to find the appropriate U.S. Geological Survey topographic maps, I drive south out of Waverly on U.S. 23 and soon pass through Wakefield. With the sun lowering behind the western treeline, I scoot quickly through this tiny community of run-down homes and trailers, satellite dishes and old cars in yards. I turn onto Salt Creek Road, then onto the gravel drive that cuts northeast and north up the side of a hill. This is the only road that dead-ends into the area that Calvin Southworth owned in 1900. On my 1910 property map there is a habitation marked at that dead-end spot I now head toward—a building that might have belonged to Orlando or Calvin. Another road that in 1910 ac-

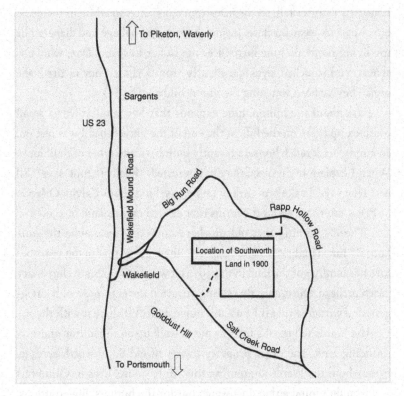

Map showing Southworth land in 1900.

cessed the land from the northwest has mostly disappeared, according to recent maps.

Excited, I varoom uphill, tires spitting gravel, and I slow only when I see an orange-vested, gun-toting deer hunter. I stop and ask permission to continue on, but permission is not his to give. He points down the hill and tells me to head back to a white house where the gravel drive begins. Down there, he says, I can talk to the Wards.

"I thought you were a census taker," June Ward tells me on her front porch. Dusk gathers round the worn gray outbuildings and barn. The house, two stories tall, seems neat from a distance, so white in the dusk, but shows its wear up close. Inside, the smell of old age and diapers fills the living room air. June introduces me to her husband, Don, who, unshaven and slouched, eyes me silently from a chair. They're tired, she says. They've been watching the grandchild all day.

I ask about the hilltop. June explains that her family owns a small complex up there on the hill, at the end of the gravel road I was just on: an empty brick ranch house, a recently built barn and other outbuildings. When I explain my project, they're interested. "Really?" June says. "I'll be," Don says. I ask them if they know anything about Calvin, Orlando or Press, anything about the farms that existed on this land in 1900.

"There's nothing that old up there now—except maybe the outhouse!" June laughs. "It's falling apart." She tells me that in the years before her family got the land in the 1930s it was an Ohio State University peach orchard. Once, an active spring attracted the enterprise of bootleggers, and we agree that if I find any moonshine, I'll share it with them.

The Wards tell me it's fine for me to walk up on the hilltop and surrounding land. The other property owner of old Southworth acreage, from whom the Wards are renting this white house, lives in Columbus and wouldn't mind either. Just watch out for the hunters, June cautions. My visit happens to coincide with the opening of deer season, and I didn't bring my blaze-orange vest. "Them deer hunters shot a chained goat once," Don says. "Thought it was a deer. They'll shoot you, too, if you're not careful."

At 5:48 P.M. on December 1, 1998, I'm alone on the hilltop that Calvin Southworth once owned. The Wards' ranch house, their barn with its concrete block walls, along with the other outbuildings, all remind me where I really am in time. But as I gaze east and southeast, I know I'm looking out to where most of the Southworth land extended. It's nearly dark, so I can't step safely into the wide meadows beyond the lawn and the gravel drive. I can still hear gunfire in the distance.

Somewhere on these acres—perhaps on this very hilltop, so wide and level, perhaps just a few yards west and northwest, perhaps in that eastern pasture behind which hills rise or maybe there in the southeastern meadow marked in the near distance by a line of trees—Buttons died. I imagine her flying over Golddust Hill, crossing over Salt Creek and landing by Orlando's barn with the Mourning Doves she kept close to for companionship. Even if the trees, the plants, the grasses have changed, even if the Southworth buildings have vanished, the slopes and curves of the land remain and nourish attentiveness at persistence, a quiet amazement at how much we can lose and forget.

A story Mary told me comes to mind. During the 1913 flood in Columbus Press Southworth and his brother Billy paddled in a homemade boat from roof to roof, rescuing people into their makeshift canoe. The waters filled Press's house, taking in their dirty currents many of his belongings—clothes, photographs, books, heirlooms, including, Mary said, the family Bible, the one "with all the history in it."

I stand by the Wards' house, beneath a security light that glares across abandoned cars and a tractor, sheds, boards and tree skeletons. All emphatically here. A bit of pigeon-leg pink colors the western sky over U.S. 23. The moon, close to full, rises in the east, while Jupiter and Saturn shine along the long ecliptic like vivid shot. The meadows' tan grasses and goldenrod do not move. No breeze blows. A rusted crop bin reads, "McCurdy," in thin white letters. Saplings grow up around an open, pitched-roof wooden shelter, just west of the house. A black satellite dish, beside which I have parked, gathers no signals.

The cold air turns blue when I step away from the harsh illumination of the security light. To the north of this hill, the lights of the D.O.E. facility—the "A-plant," locals call it—burn golden and city-like. Steam rises in columns from the plant, menacing, I think, and a water tower blinks a steady reassurance in the distance. The moon's glow softens the eastern pasture that stretches far and wide to fuzzy hills on the horizon. Power lines and their stanchions—big, metal-girdered ones—march past me from the southeastern meadow, and a V-shaped power pole rises in the foreground of my northern vista. Traffic on U.S. 23 hums, and I hear the sputter of electricity in the lines, as if a force of nature were talking to itself. Above all this, Andromeda barely shows its faint galactic green.

I get in my car, head down the gravel road and back toward Wakefield. Old farmhouses, trailers and some places I'd call shacks sit tucked along Salt Creek Road as fields give way to hilly woods that rise steeply. Christmas lights gleam off the chrome of abandoned vehicles. This is Appalachia at the fringe of homogeneous America, a place where motels fill up on the first day of deer season with Columbus executives and Portsmouth mechanics. People drive pickups here, not sport-utility vehicles, and when they get home from work at the A-plant they have a choice of not one, but two country music channels on cable. Strip-malls, fast-food joints and national discount stores line the U.S. 23 corridor, but locally owned motels and diners that once thrived bravely hang on. One advertises with a flashing sign, "Give Us a Try." And in Piketon, in Waverly, in Wakefield, you still can hear a sound of the past. Trains rumble by all night long, and their low-toned whistles wake visitors years-removed from that comforting commotion, those sounds that Press Clay Southworth felt in his bones for decades.

When I wake up at my motel in Piketon the next day, the sun shines brightly. Driving back along Salt Creek Road, I see how many of the homes seem somehow neater and more attractive in the crisp, clear morning than they had last evening. This valley was called "wild and romantic" in the 1884 *History of Lower Scioto Valley*, but today seems simply pastoral. The *History* rhapsodized further about the area's appearance in the late nineteenth century, noting:

> . . . the grandeur of its bold bluffs, the rugged outlines of its massive ranges of hills, of its dark, deep and gloomy gorges, its little valleys that here and there admit the shimmering rays of the glorious sunlight [which] makes a picture the traveler drinks in with silent awe. Then again as a dark cloud obscures the sun's bright rays, a weird and ominous-like

gloom pervades and hovers over its wild and mystic water course, giving shape to the imagination of phantom spirits reveling in the spirit world.

I spy no phantom spirits in the sharp sunlight, though I may have felt them last night. The Salt Creek valley conveys to me, as I drive, not so much a grandeur of place as the grandeur and relentlessness of time.

Back by the hilltop house and barn, I find that the only revelers today are six hunters, talking and laughing, happy to be off work and out for a morning. One of them, a friendly and burly A-plant employee, loans me an orange hat and tells me to stay near the house and not venture to the surrounding meadows. Guns fire to the south and northeast. A dozen juncos fly from a cedar tree to a brushpile, while others pick at the gravel drive. Chickadees call. I stroll through dewy grass around a decrepit pickup and a rusting camping trailer. Vines grow up around a neglected pontoon boat. This hilltop gathers to itself only recent detritus—a lawn

A view to the south, with satellite dish, from a hilltop once owned by Calvin Southworth, father of Orlando and grandfather of Press Clay. All the land in view was Southworth property in 1900.

mower, a crumpled compact car—nothing a century old. I gather burst milkweed pods and fallen oak leaves to take home.

"Are you a biologist?" one of the hunters asks as they get ready to step into the meadow sloping down toward the north.

"No." I pause. "I'm a historian." I explain my interest, say a little about Passenger Pigeons, and the men nod quietly. "The ones that carried messages?"

One of the hunters, a man in his late twenties, has the slurred speech of the slightly retarded, as he describes raising tumblers, a form of domestic show pigeon, who do exactly what their name indicates. "They were pretty," he says, "falling out of the sky like that."

On the east side of the Wards' barn, I'm alone and look toward wet, sunlit cattails, just where the pond is marked on my topo map. I startle away a Northern Mockingbird, which flies past a small flock of Eastern Bluebirds perched on the phone line.

As I walk toward a trash pile beside a locust tree, I think of the first humans to appear in this area, around 800 B.C. The lower Scioto River region was important ground for prehistoric Native Americans, including the Hopewell, famous for building massive earthenwork mounds. Later tribes, such as the Shawnee and the Delaware, settled here. Researchers have found *indigenous* trash in this county. In the eighteenth century, only a few thousand Indians lived in what is today the entire state of Ohio, but by 1844 all the tribes of this area were gone. Of the wild pigeons, the 1884 *History of Lower Scioto Valley* said, "This family have now deserted the valley for homes more retired." I discover a rusting metal sign that declares, "Landmark."

Beneath some tulip poplars beside the Wards' empty ranch house, I take off my parka, put away my camera and greet the hunters walking back from the northern pasture. I have noted approvingly how well-focused they are on the necessity for frequent breaks, to indulge in conversation, beef jerky, coffee and cigarettes. I return the orange hat to the A-plant worker and drive back down the hill past oaks and pines, where,

*Looking west from the hilltop. Calvin Southworth owned some land
just west of this point, including all that is seen in the photo.*

perhaps, an endangered timber rattlesnake still makes its home in dark
passages of rock and earth.

At the entrance to Salt Creek Road I turn left to make a circuit of the
area. Where the powerline crosses the road, down from the higher
meadow I was just near, a beautiful fenced prairie exudes such golden
light I stop for several minutes, despite the barking of vigilant dogs.
Homes with neat, wide lawns show along the north side of the road, and
cows graze under metal power poles. The hills behind those houses lead
toward Southworth land.

I take a hard left onto Rapp Hollow Road and stop at a small house,
behind which a dirt drive heads south and west toward the northern edge
of Calvin's property. A plump, middle-aged woman answers the door, letting out a blare of TV game show. She's renting the house. The land at the
end of the dirt road belongs to the Crabtrees, she says, smiling, and, no,
they wouldn't mind my heading up there.

So I drive my rattling car ever so slowly up a deep-pitted road punctuated every few inches with tire-menacing rocks. After a few slow minutes, I stop beside a fence lined with arching boughs of a shrub bearing a red fruit that I wish I could name. Around me billow green pastures. To the west and south of this point was Southworth land. I watch mallards on a pond near the Crabtree cabin, then look across a cow gate at the end of the dirt road. Leaning on one of six round hay bales where a pasture begins, I stare at a pile of shingles, in the same place, approximately, as a habitation marked on my topo map. But the fence deters me from crossing over to investigate.

Wind rustles the red-berry bramble all tangled with barbed wire. A sulphur butterfly, as if transported from the Tensas River, flies above a pool of water made by tire track and rain. Pasture, pond, cabin, sky. I like this place, I decide, and wonder how Orlando, Maria and their children felt when they had to leave these hills, when again they uprooted and moved on.

Just before I go, I investigate a gray, low-slung wooden structure—a collapsed chicken coop?—and consider whether I should pry a nail loose to show to an antiques dealer, someone who might tell me how old it is. I make the mistake of looking at my watch. I need to go.

Along Rapp Hollow Road, my car's tires humming on smooth asphalt, I journey toward home, wishing obsessively that I had taken a nail. As I do, I hear the voice of Buttons, which seems, unreasonably, I know, like a rebuke for my failure to gather that single artifact.

Buttons. She spoke once, in a human voice, and that is what I conjure as I drive. Almost 50 years ago, the world's last wild Passenger Pigeon was given human language on a radio show called "Once Upon a Time in Ohio." The 15-minute episode, titled "The Passenger Pigeon—A Lesson in Conservation," aired on Columbus's WOSU at 1:30 P.M. on November 15, 1949. The cast for the radio play included pioneers, Indians and pigeon hunters, each presenting a vignette dramatizing their relationships to the wild pigeon. Another character, a man named Mr. Holton—who knows if he was real?—described his tour of Ohio with the mounted Buttons, as he visited schoolchildren to lecture about the extinct species. After Holton's talk, Buttons concludes with a cheerfully earnest and eerie speech whose last line repeats over and over in my mind:

> See? I told you I was important and that I could tell quite a story. I'm back on my shelf at the Museum now. And I like it very much. I feel that I have a mission in life. The door is opening now and I believe some children are coming in to look at me. I hope lots of children come here to the Museum to see me because I like children and I know that when they look at me they think about conservation and all it means to our country. Do come and see me. I'll be watching for you.

Martha's Story

THE CATERPILLARS INVADED CINCINNATI IN 1872. LEGIONS OF THEM methodically munched the leaves of oak, beech, maple, chestnut, elm, sycamore and hornbeam growing along graceful streets and rough trails. They devoured the leaves of trees shading steep hills and hollows above the "Queen City" beside the wide Ohio River. People worried: Would the trees be stripped bare, would the lovely forest be destroyed?

Wasting no time, Andrew Erkenbrecher, a businessman and civic leader, raised $5,000 to organize the Society for the Acclimatization of Birds, then waited—no doubt restlessly—as a fellow Society member scoured Europe for songbirds with a taste for bugs. Armin Tenner returned with a trans-Atlantic cargo of 1,000 birds, including nightingales, European Starlings and House Sparrows. Most of them were freed in May 1873 to wage war on any swarms of caterpillars that remained or might appear again.

How many of those European birds survived—and how the caterpillars fared—are details unmentioned in the few accounts I've seen. Cincinnati still has trees, starlings and sparrows, even if it lacks nightingales.

A month after the bird releases, Erkenbrecher and others began to raise stock for a Cincinnati zoological garden. It opened to the public in 1875, second in America only to the Philadelphia Zoo, which had begun operation just months before. The remaining exotic birds that Erkenbrecher hadn't released in the anti-caterpillar campaign became part of the new Cincinnati Zoo, which also featured a tiger, an unruly elephant and one blind hyena. It was not an auspicious collection. (Later, as the zoo achieved more prominence and a sense of flair, it exhibited the rare Whooping Crane, as well as a "village" of live Sioux Indians.)

Had it not been for the caterpillars, there might not have been a Cincinnati Zoo for years or decades to come. Had the Society for the Acclimatization of Birds not transformed itself into a zoo, the story of Martha, the world's last Passenger Pigeon, would have been vastly different. Probably she never would have been on public display in Cincinnati, where, for years, she was the zoo's most famous attraction. Perhaps some other Passenger Pigeon would have been celebrated, then largely forgotten, as the last representative of this astonishing species.

So we have remembered, fitfully and often inaccurately, the lonely, lovely Martha. Many visitors to the Smithsonian Institution in Washington, D.C., have seen her stuffed specimen on display there. Birders who know little of Passenger Pigeons often know the name of Martha. But most of us know nothing of her life and death—and her strange life after death.

The history of Passenger Pigeons at the Cincinnati Zoo begins in 1874, when one Frank Louck, Esq., of St. Bernard, Ohio, donated two Passenger Pigeons to the nascent facility. Those birds may have seen a few wild pigeons feeding and roosting in nearby trees and may have watched the last flocks of their kin fly above Cincinnati in 1876.

A few years later another Passenger Pigeon would be born and be named for America's first First Lady, Martha Washington. Sometime in 1902, 1900, 1897, 1896, 1895, 1894, 1889, 1888, 1887, 1886 or 1885, inside an egg, a blind, cramped, featherless female Passenger Pigeon tucked her bill between her body and a wing. Pecking with her bill at membrane and shell for several hours, she began to fracture the white egg that had kept her safe. The temporary egg tooth on her bill helped with the ordeal; it would vanish after she hatched out. Assisted by an extra muscle on the back of her neck, which soon would become vestigial, the chick finally broke off a tiny fleck of shell. A splinter of light touched her skin for the first time. She squirmed and pecked. The egg fractured into wider cracks, until the chick stretched her neck upward. Born into a caged life, Martha might have enjoyed a taste of wildness: the parent birds feeding her their rich pigeon milk.

The year of Martha's birth remains uncertain because the man who ran the zoo—Sol Stephan, an ex-circus elephant trainer—changed his stories. Stephan and his son, Joseph, both gave multiple versions of Martha's birth. Sometimes they claimed to have purchased Martha; often they said she had been born at the zoo. So convoluted is this history that different female Passenger Pigeons at the zoo may have been called Martha at different times. Faulty memory, bad recordkeeping and a desire for enhancing the zoo's reputation seem the most likely reasons for all the conflicting accounts. Perhaps the Stephans felt it was more compelling to claim that the world's last Passenger Pigeon had been born in the facility they supervised. Sol also may have been given differing accounts of Martha's origin by the keepers, who were more familiar with the zoo's birds on a day-to-day basis. Unfortunately, files that might have shed light on Martha's origins were destroyed by a 1963 fire in the zoo's administration building.

We do know that in 1878, the zoo purchased some pigeons at $2.50 per pair. Some of these birds are known to have bred. According to an annual report, two Passenger Pigeons were born at the zoo on June 26, 1889. Was *that* Martha's birthday?

We also know that in fall 1888, a Milwaukee pigeon breeder named David Whittaker either purchased from or was given by a Native American four Wisconsin Passenger Pigeons. These birds had been captured close to Lake Shawano, just northwest of Green Bay. One died and one escaped, but the remaining two successfully mated. (Whittaker kept his birds in an outdoor cage near his house on a hill above the Milwaukee River where, when storms approached, the pigeons sat beside each other on a perch, tucked their heads in, then soon spread tails, stretched wings and flew up against the cage, wanting to escape.) University of Chicago professor Charles Whitman bought seven Passenger Pigeons from Whittaker in the late 1890s. From Whitman's group, apparently, the Cincinnati Zoo obtained Martha—or a Martha—in 1902. (That same year two pigeons escaped from Whitman's Woods Hole aviary.) It's worth noting that Sol Stephan explicitly wrote in 1907 that the zoo's last female pigeon was received from Whitman in 1902.

Though she never flew free, Martha did cross the country in Professor Whitman's care, as he took his pigeons from Chicago to Woods Hole, Massachusetts, and back again on the New York Central Railroad, according to pigeon researcher and fancier Joseph Quinn. Whitman kept research aviaries in both locations. Today a handball court beside brick apartments occupies the space where Whitman's Woods Hole aviary once stood.

By early 1909, the birds owned by Whittaker and Whitman had died, and just three pigeons, including Martha, remained at the Cincinnati Zoo. Had Whitman known earlier that Passenger Pigeons became more robust after eating earthworms, his flock, he believed, would have persisted.

Desperate attempts to find other mates for the zoo's pigeons came too late. There had never been a cooperative effort to captive-breed the Cincinnati birds when Passenger Pigeons were alive at various aviaries. At one point, the zoo even had a group of 20 Passenger Pigeons, which, if bred with others owned by different facilities and aviculturists, might have increased genetic diversity just enough to have saved the species. Perhaps there just weren't enough birds left. Hybrids that had been produced by breeding Passenger Pigeons with separate species turned out to be sterile.

Then, on an unspecified day in April 1909, one of the zoo's two male birds died, leaving only Martha and George (named, obviously, for George Washington). The last known pair of Passenger Pigeons anywhere.

Of Martha's life in Cincinnati we know relatively little, at least as concerns the textures of her days and nights. We don't know how often a bird keeper placed a tray of grain or nuts or earthworms beside Martha in her cage. She lived a mostly sedentary existence and watched visitors watching her and George, as the birds called to each other a sound that writers translated as "See? See?" From time to time, the pigeons sidled along branches and perches—a mark of mating behavior—and occasionally constructed loose, twiggy nests in a tree confined within their cage.

George died on July 10, 1910. His body was not preserved because the plumage was in a "poor state," according to one naturalist. Martha was now alone. Probably she could not have missed George in what we'd call memory, though who can say? Certainly she lived in the insistent alone-

Martha in her outdoor cage. The photo was taken either by Enno Meyer or Dr. William C. Herman. This may be its first appearance in print. The photo may be an unretouched original by Herman; if so, it probably is the last photo of Martha while she was alive. The date is unknown.

ness of each moment after George's death. Her eyes saw cages, concrete, seeds, passing flits of birds beyond. She saw none of her kind, and this may have been a kind of anxiety in a bird as intensely social as the Passenger Pigeon.

Joseph Stephan, son of zoo General Manager Sol, had sometimes ferried Passenger Pigeon eggs from the zoo to his home, letting domesticated pigeons warm them. When the eggs were ready to hatch, Joseph returned them to the rightful parents, who, he wrote, "mothered them O.K." Now he watched the only pigeon left. There must have been days when Joseph wondered how long Martha would last, days when he walked by the elegant Zoo Club restaurant with its wide wraparound porch and columns, just across from the Band Stand. Leaning against a tree, he might have watched the meanderings of swans and pelicans on the lake or the gondola ferrying a courting couple.

From time to time, Martha flirted and stretched her wings, missing the flight she never really had. Mostly she sat. She ate at feeding time and stared beyond the metal mesh.

By the time it became clear that Martha was the last Passenger Pigeon, people were taking notice. The rival New York Zoo desperately wanted Martha, but Stephan refused to sell her. She would stay in Cincinnati surrounded by hills where her ancestors had once flocked and where the first trapshooting club in the United States had originated, using live Passenger Pigeons as targets.

Martha attracted a bevy of visitors to her home in one of the seven stone pagoda-like aviaries—the Bird Run—nestled among shade trees across from the Monkey House, the Pheasant Yard and the Carnivora Building. Her aviary measured 18 feet by 20 feet, and a metal "summer cage" about doubled her living space. The aviary's tin roof, shaped to look like tiles, was almost as red as Martha's eyes.

On the aviary, Sol Stephan placed a sign, now lost, which declared that Martha was the very last Passenger Pigeon. "There were many scoffers," Stephan recalled. People had a hard time believing the pigeons were gone, all but this one. Ornithologists and naturalists didn't scoff. They arrived from faraway cities on pilgrimages to see Martha. For the first time, a creature about to be extinct had become a celebrity, subject to a bleak, surreal fame.

As Martha grew older, she might have stepped or flown forward each time Joseph arrived. Perhaps she ate directly from his hand. But, otherwise, she stayed motionless, which irritated some of Martha's curious visitors. Joseph remembered that "on Sundays we would rope off the cage to keep the public from throwing sand at her to make her walk around . . ."

One newspaper reporter wrote that Martha "is to be seen in the open air cage opposite the entrance to the lion house. There will be no mistaking the bird, as its drooping wings, atremble with the palsy of extreme old age, and the white feathers in the tail, make [her] a conspicuous object." Martha steadily weakened over the years, so the bird keepers lowered her perch until, finally, she could only stand on the ground, literally dragging her long tail.

"When the bird started to mope and show other signs of approaching dissolution," Stephan remembers, "I felt personal grief." Not only had Stephan cared for Martha, he had witnessed wild pigeon flocks, the bloody business of market hunting and pigeons caged for a trapshooting contest. He understood what had led to her singular status, even if, because of the times in which Stephan lived, he and others had failed miserably to imagine how the birds could have been saved.

As Martha molted in the summer of 1914, Sol and his keepers collected her fallen feathers in a cigar box. They knew that whoever mounted Martha's body after she died would need to glue the feathers back on her skin.

"The days of the last passenger pigeon . . . are now numbered," an anonymous reporter for the *Cincinnati Enquirer* wrote on Tuesday, August 18, 1914. These words appeared beneath Martha's picture:

It has lived for almost 30 years at the Cincinnati Zoological Garden under the tenderest care of General Manager Sol A. Stephan, but he has abandoned hope of keeping it alive more than a few weeks longer at the very most. That it has been failing rapidly has been noted for some time, but it was not considered more than the feebleness of extreme old age until yesterday morning, when Superintendent Stephan discovered it early in the morning lying on its back apparently dead. A few small grains of sand tossed upon it shocked it into activity again, and last night it was acting stronger and fed heartily when the evening feed was offered.

The article went on to explore the causes of this impending extinction, including the "wholesale destruction" of Martha's ancestors by market hunters, as well as the conjectured possibilities of sudden diseases and accidental mass drownings.

Sol Stephan awoke on the late-summer days of 1914 wondering each morning if Martha had made it through the night. On Tuesday, September 1, 1914, Martha roused from a warm night's sleep; the early morning low had reached just 72 degrees. History does not record the name of her first human visitor. But the *Cincinnati Enquirer* for that day does record news of the Great War in black headlines of portent and carnage. A small poem on page 4 exclaimed, "All hail September. Hear our praise / Oh, month just newly born!" On the bottom of the same page ran a news item concerning an upcoming meeting in Winnipeg of American and Canadian lumbermen. By 1914, logging—implicated in the extinctions of the Carolina Parakeet, the Passenger Pigeon and the Ivory-billed Woodpecker—had become a big enough business that its practitioners organized conventions.

Tuesday's advertisements caught the attention of readers: From 2:30 to 8:15, Weber and his Band would play dance music on the Zoo Clubhouse porch; at Rollman and Sons, the new fall hats had arrived, including those decorated with the feathers of birds; and the New Arlington

Hotel promoted itself as an "attractive summer resort [with] ... Universal relief from hay fever." The hotel, coincidentally, was located in Petoskey, Michigan, near the site of the last great pigeon nesting in 1878, where H. B. Roney and his comrades had tried to bring attention to the illegal and devastating slaughters there.

On Tuesday, September 1, 1914, travelers left Cincinnati for other locales via the river boats Chilo, Greendale, Greenland and the City of Cincinnati, all plying the mighty Ohio River. A world away, the Turkish Army mobilized in support of the Huns. Just a few counties distant, in Waverly, Ohio, near where the last wild Passenger Pigeon had been shot 14 years before, a new school superintendent pondered the semester's impending work. And later that afternoon, under partly cloudy skies, the Cincinnati Reds would host the Chicago Cubs, losing to the visiting team 8–7. All the players would be drenched in sweat, the humidity reaching 74 percent, the temperature rising to 89 degrees.

While we have a record of all these details, the particulars of Martha's death remain indeterminate and conflicting. Martha died alone, probably at or about 1 P.M. on Tuesday, September 1, 1914, as recorded by Schorger. This is the time and date now generally accepted. Yet, according to Joseph Stephan, "Martha died at 5:00 P.M. September 1, 1914. My father ... and myself were with her at her death ..." According to the *Cincinnati Times-Star* and *Field & Stream*, Martha "was found dead shortly after noon by William Bruntz, its keeper, beside [her] low roost ..." Sol Stephan wrote to a Pennsylvania taxidermist not long after Martha's death and said that she had died at 2 P.M. on August 29, but nothing else supports this assertion. Everyone agrees that the image of Martha taking her last breath while surrounded by stunned and grief-stricken ornithologists is just a fanciful legend.

Whoever first found Martha probably tossed grains of seed or sand or gravel onto her ragged feathers, hoping she'd awaken. She didn't, and thus the titanic vanishing of the Passenger Pigeon concluded, finally, on the bottom of a cage in the middle of a city busy with commerce and worry about war.

Eugene Swope, the Ohio representative of the National Audubon Societies, telegraphed Audubon leader T. Gilbert Pearson to tell him what had happened, provoking Pearson to comment—in the words of one newspaper—that Martha's death "is a calamity of as great importance in the eyes of naturalists as the death of a kaiser to Germans throughout the world."

Sometime on September 1, an anonymous writer at the *Cincinnati Enquirer* typed the obituary, reporting Martha's age at 29 and noting that:

> There will be no funeral for Martha. Instead, her remains, together with the feathers that she has shed in moulting, will be shipped to Washington, to be preserved in the Smithsonian Institution. Martha will remain to be shown to posterity, not as an old bird with most of her plumage gone, as she now is, but as the queenly young passenger pigeon that delighted thousands of bird and nature lovers at the Zoo during the past 20 years.

To keep the body from rotting in the heat as it traveled by train from Cincinnati to Washington, D.C., Joseph Stephan "took her to the Cincinnati Ice Co. plant personally," he wrote, "and supervised the placing of her body in a tank of water, suspended by her two legs, and froze her body into a 300-lb. block of ice." Another version of the story has Joseph Stephan cutting holes into existing ice blocks to hold Martha's body. Pigeon fancier Joseph Quinn maintains that Sol himself did the ice-cutting.

A photograph of Martha "on ice" raises some questions about how her body was prepared. The picture probably could not have been taken in Cincinnati if she had been lowered into water and then frozen into place, unless the ice was very clear. If the summer sun melted all the ice

Martha on ice, after her death. Note the faintly visible incorrect date of "1915." Why the date occurs and who wrote it are mysteries.

during the long train trip, the photo could not have been taken at the Smithsonian. The most likely scenario is that Joseph did in fact take Martha to the Cincinnati Ice Company, where she was slowly frozen in place, and that enough ice lasted the journey to Washington for this photograph to have been taken there.

A recital to have been presented at the Cincinnati Zoo on the day that Martha died was to have included Francesco Paolo Tosti's musical setting of George John Whyte-Melville's poem "Goodbye"—a grimly appropriate farewell.

> *Hush, a voice from the far away.*
> *Listen and learn it seems to say.*
> *All tomorrows shall be as today.*
> *The cord is frayed, the cruse is dry.*
> *The link must break and the lamp must die.*
> *Goodbye to hope, goodbye, goodbye.*

As the singer sang, the Stephans said their own farewells, whatever they were, to one of the most famous birds in the world.

On the morning of September 4, 1914, Dr. Charles Richmond, assistant curator of the Smithsonian's Division of Birds, called R. W. Shufeldt at his home. Richmond told Shufeldt, a physician, ornithologist and museum associate in zoology, that officials wanted the respected naturalist to perform Martha's autopsy. Martha had arrived at midmorning, with the Smithsonian paying 54 cents for a delivery-due charge. An accession memorandum recorded the event, in Dr. Richmond's handwriting: "1 Passenger Pigeon *(Ectopistes migratorius)* in the flesh. The death of this individual marks the complete extinction of the genus and species. . . . "

Shufeldt rushed to the museum and had the bird taken to a studio where he and Smithsonian photographer T. W. Smillie took a number of exposures. Although some tail feathers were missing and some feathers were soiled, most of Martha's plumage "was smooth and good." Martha, Shufeldt wrote, "had the appearance of a specimen in health . . ."

Then, at 1:15, Shufeldt and assistant taxidermist William Palmer left the museum, with Palmer carrying the bird. The men headed back to Shufeldt's house, where Shufeldt could develop negatives and Palmer could "skin" the bird in a workroom. Throughout the afternoon, Shufeldt moved from the darkroom to Palmer's desk, taking more photos of Martha. They are gruesome pictures, images of exposed sinews and organs. Palmer placed Martha's brain and eyes in a jar of alcohol; by day's end all of her internal organs and tissue—minus skin and feathers— would be preserved in such a jar.

After a late lunch, Palmer left with Martha's skin and feathers. Once the Smithsonian's chief taxidermist, Nelson R. Wood, returned to town, he would mount Martha in a lifelike pose and the bird would go on display. Left alone with Martha's skinless body, Shufeldt lifted it and prodded. He looked carefully, discovering an inexplicable cut in the abdominal cavity as well as much damage to part of her liver and intestines; they were both disintegrated—"as though it had been done with some instrument." Further, many of Martha's organs were not where they were supposed to be, but Shufeldt offered no public speculation for the cause of this rearrangement. He picked up his scalpel and set to work, opening this or that chamber, finding atrophy in, for example, the ovary. The lungs looked dark; apparently, Martha had suffered recently from lung congestion.

Reading Shufeldt's autopsy notes, one confronts the probing of both a body and a spirit. Meditating on how people would react to Martha's display at the museum, Shufeldt envisioned much interest but little expectation that it would serve to stave off extinctions yet to arrive. "In due course, the day will come when practically all of the world's avifauna will have become utterly extinct," he wrote. "Such a fate for it is coming to pass now, with far greater rapidity than most people realize." During the autopsy, Shufeldt made a decision based not on science but on sentiment.

He "did not further dissect the heart, preferring to preserve it in its entirety . . . as the heart of the last 'Blue Pigeon' that the world will ever see alive."

Martha's heart stayed whole, and her status grew. And as years passed, more and more Americans who visited her at the Smithsonian understood that there were, in fact, no more Passenger Pigeons. The rumors and "sightings" of the early part of the century ceased. The unalterable fact of extinction—as a kind of abstraction—became clearer once one looked at Martha's stuffed body. So, inevitably, Martha accrued a certain symbolic power: Martha represented the finality of extinction and the consequences of the failure to conserve our natural resources.

In 1939, for example, organizers of a sportsman's show in Cincinnati tried to get Martha returned for the 25th anniversary of her death. They wanted her mute body to testify to those broad lessons. (The Smithsonian refused to part, even briefly, with such a valuable specimen.) In 1942, the chief biologist for the Soil Conservation Service wrote to the Cincinnati Zoo, hoping to obtain newspaper clippings about Martha, because he saw an analogy between the decline of the Passenger Pigeon and the loss of topsoil; he wanted to make the parallel clear in soil conservation materials for farmers. Martha's death was, and remains, a kind of ready-made lesson, a parable for other possible conservation tragedies. Ironically, this attention to generalities, to "meanings" about Martha, also helped create subtle obstacles to understanding *the details* of her life and her species.

Because she was a symbol, Martha was able to take her first extended flight. She left Washington in 1966 to travel to San Diego, where the San Diego Zoological Society Golden Jubilee Conservation Conference convened in early October. Zoo officials there had told the Smithsonian that they wanted "something comparable in a way . . . to the display in this country of the Mona Lisa . . ." The Smithsonian's Division of Birds Assistant Curator George Watson agreed to send Martha so long as she was covered by a $5,000 insurance policy from Lloyd's of London and would be displayed only in a low-light setting, to avoid damaging the feathers.

Technician Ted Bober carefully prepared a box for the specimen and took the bird to the airport. Remarking on Martha's first flight, a reporter for the *Washington Evening Star* joked, "There is a question as to whether she should be listed as a passenger pigeon or a pigeon passenger." Curator Watson wasn't interested in puns. He was too concerned about the possibility of damage or loss. So he specified that Martha should "sit on [a stewardess's] lap all the way." A news photo shows "one of American [Airlines's] prettiest," Miss Nancy Evers, "who struck up an immediate attachment for her feathered friend." If Evers didn't keep Martha's 15-pound box on her lap during the entire trip on September 19, 1966, she did keep guard over the specimen. And so Martha traveled first class to the edge of the Pacific.

After the conference, she returned on October 18, five days *after* the insurance policy had expired. But Martha was safe. Rather than deposit her in storage—where, years before, she had been misplaced—technicians carefully put Martha back on display. Her curators believed that's where she would remain.

But the Smithsonian hadn't counted on a respected wildlife painter named John Ruthven, who, one day in 1973, got some upsetting news: The old "Bird Run" at the Cincinnati Zoo was to be destroyed. The seven graceful aviaries, among the first buildings of the zoo, were to be razed for much-needed renovations to the ape facilities. Even the aviary that had housed Martha and the world's last Carolina Parakeet, Incas, was slated for demolition. Ruthven found the prospect repellent.

Ruthven's love for birds and woods was colored by the fact that Martha had died just 10 years before his birth. "I didn't get to see her, and I regretted that," Ruthven remembers. So when the zoo's director, Ed Maruska, told him it would cost $50,000 to move the aviary 50 feet in order to preserve it, Ruthven offered to do a print of Passenger Pigeons and donate funds from the sales to that very cause. He raised $40,000, but an anonymous donor suddenly paid for the building's relocation. The $40,000 from Ruthven's sales were applied to renovation and remodeling.

The artist's activism wasn't the only reason the zoo agreed to save the aviary, according to Maruska. He tells the story of the Zurich zoo director who traveled to Cincinnati sometime in the early 1970s, specifically to see where the last Passenger Pigeon had died. The professor expected at least a plaque. But there was nothing, no marker, no sign. Maruska felt deeply embarrassed. Now, by saving the aviary, the zoo would be saving history. And, as Maruska says, the whole incident reinforced a feeling about another kind of saving: "We made a vow that we'd never let a species go extinct at the Zoo again."

Ruthven's effort culminated in his winning permission from the prominent Smithsonian ornithologist Alexander Wetmore to let Martha out of her glass case one more time. Martha would fly home to Cincinnati. Her display would be the centerpiece of a ceremony formally launching public fund-raising for the aviary renovation and for educational programs.

On a June day in 1974, staffers with the Smithsonian's Division of Birds again removed Martha from her Bird Hall display. Technician Phil Angle took pictures as Martha was placed carefully—still on her wooden branch—in a plastic bag that nestled amid Styrofoam peanuts in a simple, sturdy cardboard box.

Ruthven asked Betsy Nolan, his New York–based publicity agent, to escort Martha to Cincinnati. Handed the box labeled "Martha: The World's Last Passenger Pigeon" (with contents insured, as they had been for the San Diego trip), Nolan took the specimen to National Airport and presented two first-class tickets, one for her and one for Martha. When American Airlines Flight 275 became airborne a few minutes after noon on Thursday, June 27, 1974, Martha was back in the sky. The flight crew even announced her presence on the plane.

She arrived in Cincinnati at 1:15 P.M. on what the mayor and gover-

Betsy Nolan and John Ruthven unpack Martha.

nor had designated as Passenger Pigeon Day. Betsy Nolan looked for her bags, which never arrived. But at least the World's Last Passenger Pigeon would not spend the weekend in the backroom of some airport's lost-luggage depot.

For the next three days, this "feathered conscience," as Ruthven called her, appeared on display from 9 A.M. to 6 P.M. at the zoo's Reptile House. Bill Mers and Richard Fluke, who had seen Martha when she was alive, looked at her now-diminished colors. They remembered her red eyes blinking, her neck curving down as she pecked at seed. Mers recalled the sign that told of the zoo's never-claimed reward for discovering a mate for

Martha. Fluke, a former fur trapper, said he had taken up nature photography to help raise awareness about the plight of threatened species. After speeches in Martha's honor, visitors could pay their respects, then attend a show of "The Derbys and Their Affection-Trained Animals from Television and Movies."

When the weekend ended, Nolan took the specimen aboard American Airlines Flight 400, leaving Cincinnati on Monday, July 1, 1974, at 7:50 A.M. and arriving at National Airport at two minutes after 9. By the afternoon, officials had the World's Last Passenger Pigeon ensconced in her glass case. She would never again leave the air-conditioned Smithsonian.

In Cincinnati, the renovation of the aviary continued, with the help of nature writer George Laycock, the Langdon Club (a local naturalists' group) and other artists and citizens. Cincinnati sculptor Robert McNesky unveiled his bronze statue of a Passenger Pigeon in 1976; the statue today stands at the entrance to a little path that surrounds the aviary shrine. McNesky also carved the aviary's massive wooden doors with reliefs of images of other extinct and threatened birds. The renovated building, listed on the National Register of Historic Places, opened on September 1, 1977, with another ceremony. Again, 81-year-old Richard Fluke spoke. This time he cried.

Just a week after the dedication, a burglar broke into an exhibit case at the new Passenger Pigeon Memorial and grabbed a relic, 16-gauge shotgun that Ruthven had donated. The robber escaped, cut off the barrel and later robbed a taxi and grocery store. "The assailant loaded the gun with modern shells," Ruthven tells me, laughing. "If he had fired, it would have killed *him*." Fortunately, no one was hurt, the robber was prosecuted and the gun was recovered for public display.

All these stories were in my mind one late spring day, when I finally had the opportunity to visit the Passenger Pigeon Memorial. I tugged on those dark, heavy doors, but they would not open. So I strolled around the

A view of the stony aviary in which Martha lived for many years.

small aviary and fumed in disbelief: I had driven all the way from Kansas and, now, could not get in.

Half-jogging across the grounds, I reached the office of Jan Dietrich, the zoo's school services coordinator. Waiting for her to fetch the key, I kept my distance from "Telly the Bald Ibis," who, with a long and threatening beak, watched my moves with great interest. When I returned to the aviary and Jan unlocked the door, I stepped inside—triggering an alarm, painfully loud, that drove us back in an instant. No one could steal a gun from this facility, I thought, and Jan rushed to get another staffer while I stood far from the open door. The siren blared. What I had intended to be a pilgrimage had instead become a series of absurd misadventures. I sweated and waited, smiling a forced smile while zoo personnel drove by slowly, back and forth on golf carts, and eyed me much as had Telly the Bald Ibis.

Finally, someone swooshed up the walk and turned off the alarm. But when I stepped back into the building, I couldn't find a light switch. Appropriate, I thought. I'm in the dark in Martha's home. So, with a flashlight, I peered at shadowed displays—an alligator skin belt, a net that had once trapped live pigeons. I admired Audubon prints of the Carolina Parakeet and the Passenger Pigeon, as well as actual specimens of each. And here, yes, was John Ruthven's shotgun, its barrel filled, lest anyone think the weapon could work.

I tried to imagine Martha in her cage in that very place, but couldn't. Too flustered, I felt no epiphany. I stood a moment in the dark while Jan waited outside and then I stepped, blinking, into the sunlight that filtered through trees and shrubs surrounding the limestone building. House Sparrows flitted through leaves. When two boys walked up the path and saw the shrine, one of them said, "This is where Martha and Incas died." I smiled at that small moment of recognition—not so small, really—then decided to go.

Recently, at the Smithsonian, Martha has been in temporary storage to accommodate a building renovation. Perhaps someone will clean her dingy feathers with mild soap and water before she is put back on public display.

But in addition to her static body on a shelf or behind glass, Martha has a different existence. She abides as a "3-D" image on the World Wide Web. (The website address is kept confidential, except to those who need to use it for actual research.) Having photographed Martha with sophisticated cameras that captured multiple views for a three-dimensional effect, the Smithsonian has made the famous pigeon available to scientists who otherwise might have had to touch the specimen. Even before a researcher decides to study an object—measure it with calipers, say, or take detailed notes on colors—he or she must lift it and scrutinize. Such an initial examination determines whether the item is worth closer study. It's also damaging. "Seventy percent of the wear and tear on objects comes from handling *before* someone even begins studying the piece," explains

Carl Hansen, branch chief of the Smithsonian's Center for Scientific Imaging and Photography at the National Museum of Natural History.

Martha's one-of-a-kind status made her a logical choice as an initial "3-D" photograph on the Smithsonian website, which will soon feature a wide range of other artifacts for on-line study. "She is such a classic object in the collection," says Hansen. So classic, he stresses, so important, that only the Smithsonian taxidermist handled her during the photo session. "If we had dropped her," Hansen says, "she would have broken into a million pieces, like Humpty Dumpty. Now, we know where every feather on her goes."

The Labrador Duck and the Great Auk

Overleaf (top): The male Labrador Duck as figured in the 1884 book The Water Birds of North America.

Overleaf (bottom): The Great Auk, from an illustration in The Family Christian Almanac, 1871.

The Strangest Sea

Resurrection of creatures from our planet's prehistoric past no longer seems an impossibility.

> —Rob DeSalle and David Lindley, *The Science of Jurassic Park and the Lost World, or How to Build a Dinosaur*

Extinction really is forever, in spite of what we are led to believe in dinosaur movies.

> —Ornithologist David Steadman

IN A MOTEL IN OHIO, HAVING RETURNED FROM A DAY OF RESEARCH, I SAT alone, with a sandwich for dinner and a television for company. I found myself watching, thoughtlessly, the famous velociraptors of Hollywood as they destroyed the fictive structures of their origin. At the end of *Jurassic Park,* a movie of enthralling imagery and epic silliness, there occurs a scene in which the tired protagonists escape the island home of rampaging dinosaurs. The survivors watch from a helicopter as graceful pelicans fly through the air. This is a moment of relative nuance in a movie of obvious narrative gimmickry. For birds are, most believe, the modern descendants of dinosaurs. The pelicans remind us that lives here and now carry traces of the past.

Yet that is not the movie's primary theme. Far from being a celebration of the world as it is and far from being a cautionary tale about the excesses of scientific inquiry, *Jurassic Park* extols strange human ambitions and stokes our fascination with magnificent creatures that we can imagine guiltlessly, because, after all, we did not kill them off. So the film puts its images in the service of a powerful, and very old, narrative fantasy, the dream of ultimate human control over something more fearful than a *T. rex*—death itself.

This imagined flirtation with divinity rests, of course, on our ability to suspend disbelief about the fundamental premise of the story: that we can restore dead lives through genetic manipulation. While I watched the movie—I admit, this was my third time—I could not help but wonder about the birds I have written of, these lives I have come to love. What *if* we could bring them back? What if we could release into the sky flocks of jewel-green parakeets and clouds of blue-gray pigeons? I pondered a moment the ethics of such a prospect, but soon felt distracted by the lure of the act itself. Rebirth. Resurrection. Such potent words. The question in my mind soon became not "what if?" but, perhaps indulgently, "which one?"

As the cinematic ocean disappeared on screen, replaced by a commercial, I knew the answer: the Labrador Duck. Of all the birds I have chronicled, this one remains the least understood. Not only have I wanted to see one alive—riding the swell and surf off some lonely tombolo—I have yearned to *know* its life, to know of it all I could.

The male of this species was exquisite, with his black body, white-and-black wings, white neck crossed by a collar of black and his white head topped with a narrow black patch. Beneath eyes described variously as red, hazel or yellow, there occurred a cheek patch of bristly, yellowish feathers. Some illustrations show a blue-gray eye stripe as well. The female was gray-brown, with white on the wings, a grayish throat and a light line behind her eye. Immature Labrador Ducks shared a plumage similar to the females, but of the colors of the downy young, we cannot say.

The black-and-white pattern of the male prompted many to call it the Skunk Duck. French speakers referred to it as Canard du Labrador. Newfoundlanders called it the Pie or Pied Duck and the Pied Bird (which, con-

Drawings from The Water Birds of North America, *1884, show the interesting beak structure of the Labrador Duck. Note especially the view of the male's bill (left), which best shows contours of the bill's wide tip and the lateral flanges.*

fusingly, was a name applied to other species, such as the Common Goldeneye and the Surf Scoter). Sportsman J. M. Le Moine wrote that the species was called the Pied Scaup Duck, a name unreported in recent ornithological literature.

The first word of the scientific name for the Labrador Duck, *Camptorhynchus labradorius*, derives from the Greek for "flexible beak," which indicates the bird's most remarkable feature, a large and odd-looking bill rather like that of a Steller's Eider. The bill was also similar to another duck's, the Northern Shoveler. The Labrador Duck's bill widened toward its tip and sported two soft lateral flanges that hung down from the beak. Biologists speculate that these flaps helped the bird use touch to locate its burrowed prey of small mollusks and shellfish. The duck also ate small fish and seaweed. Further analysis of isotopes in Labrador Duck bones may help us understand more about the diet of this species, according to ornithologist Glen Chilton.

The mandibles also had a high number of lamellae, which are plates or "serrations" inside the bill that function like a strainer. The lamellae helped the bird retain appropriately sized food and filter out sand and other detritus. Biologist R. I. Goudie found that, among sea ducks, the Labrador had the biggest lamellae in relation to body size, which supports an interpretation that the bird relied quite heavily on sand-buried shellfish.

Audubon's rendering of two Labrador ("Pied") Ducks. The female (left) was, in fact, a young male. The two specimens were shot by Daniel Webster on Martha's Vineyard and later presented to Audubon.

As well, the Labrador Duck's cere—a fleshy area on the upper mandible that surrounds the nostrils at the base of the beak—was so large that one ornithologist called it "swollen," though he did not speculate about the reason for its size. The Labrador Duck's bill was colorful, too—black, blue-gray and orange.

"It procures its food by diving amidst the rolling surf over sand or mud bars," wrote John James Audubon. The Labrador Duck also dabbled for food in shallows. And it greedily took to bait. "A bird-stuffer whom I knew at Camden had many fine specimens [of the Labrador Duck]," Audubon reported, "all of which he had procured by baiting fish-hooks with the common mussel, on a 'trot-line' sunk a few feet beneath the surface, but on which he never found one alive, on account of the manner in which these Ducks dive and flounder when securely hooked."

Another name for the species, the Sand Shoal Duck, accurately reflects its favored habitats, at least in winter. In that season, the duck foraged along the Atlantic coast in sandy bays, estuaries, inlets, harbors and coastal salt marshes, apparently preferring, like many waterfowl species, those places where underwater eelgrass grows. The winter range along the eastern seaboard probably extended from Nova Scotia as far south as the Chesapeake Bay and centered on Long Island. This is where collectors shot most of their specimens of the species. Audubon said that only rarely, when pressed by storms, did the Labrador Duck fly inland along rivers.

Human intervention temporarily expanded the duck's range, at least according to Le Moine, who wrote in 1863 that "the *Bulletins* of the French society describe a beautiful species of duck, recently introduced from Canada, in France; it is called the Labrador Duck." Unfortunately, Le Moine provided no further information. Given that no one else mentions the introduction of this species to Europe and given the duck's apparently specialized feeding habits, one must doubt whether it lasted very long in France, as either a wild bird or a captive. An additional confusion, Chilton points out, is that a kind of domesticated duck unrelated to the wild species was also called the Labrador Duck.

The species's breeding distribution remains shrouded in uncertainty because of vague record-keeping and the use of various vernacular names for the duck. As well, the species's small population at the time of European settlement made the duck hard to locate on its apparently remote breeding sites. The species may have bred in Greenland, but ornithologists more confidently believe that the bird nested on this side of the Atlantic as far north as the Ungava Peninsula, Labrador and in and about the Gulf of St. Lawrence. Ornithologist E. H. Forbush claimed that this duck "was common in summer on the coast of Labrador until about 1842" and that it was common on the northern shore of the Gulf of St. Lawrence and bred there, but neglected to tell us the source of this information. He also said, perhaps not very helpfully, that the duck "bred more or less locally." In 1929, *Forest and Stream* editor William Bruette declared, incredibly, that the species *bred* "as far south as the state of New Jersey." Had this been the case, we would know far more about the bird than we do, because its breeding would have been in close proximity to American naturalists. On what evidence Bruette makes his fantastic assertion one cannot deduce.

The Labrador Duck reportedly passed over Labrador, Nova Scotia and New Brunswick after breeding, but its migration paths are not well known. One account claimed the birds migrated through Pennsylvania when heading north to breed, but here, too, we cannot be sure.

Anything that can be said about the mating, nesting and breeding biology of the Labrador Duck remains conjecture based on scant evidence and on inferences made from related sea ducks, such as the Common Eider. Probably the Labrador Duck was a long-lived bird that did not breed until two to three years old. Male birds would have taken two years to reach adult plumage, and the females likely laid few eggs. (Incidentally, only nine pale olive eggs attributed to this species exist today; until genetic tests are performed, no one can be sure that they truly belong to the Labrador Duck.) The birds may have nested on small, rocky islets. Audubon claimed that his son John Woodhouse was shown the nests of

Pied Ducks at Blanc Sablon, Quebec, on July 28, 1833. The nests of grass, fir twigs and down were "placed on the top of the low tangled fir-bushes . . ." Possibly these were the nests of Surf Scoters. Audubon himself never saw a Labrador Duck or its nest while he was in Labrador. He did observe the birds on Chesapeake Bay and noted that they gathered in small flocks of between 7 and 10 birds, flocks he thought were families.

Unfortunately, no one recorded the duck's daily routine even in its wintering range along the populated American coast. One can't help but wonder why those few naturalists who did see Labrador Ducks couldn't refrain from shooting, even for a few minutes, in order to better observe the duck's behaviors.

No one described the calls of the species, though we may conjecture that they were similar to the coos of other sea ducks. We do know that when the duck flew, its wings whistled.

We can't even determine, finally, if the bird was trusting or wary of humans. Nineteenth-century gunner Nicolas Pike asserted, "They were shy and hard to approach, taking flight from the water at the least alarm, flying very rapidly." But, according to Forbush, the Labrador Duck "was so tame and confiding that it was not difficult to shoot." And, in *A Natural History of the Ducks*, John C. Phillips wrote, "We do know certainly that it was a rather stupid duck . . ." One would be inclined to trust Pike—he shot the birds, after all—but the assertion of tameness is supported by two other names for the species: Fool Bird and Fool Duck.

What nearly all the historical sources agree upon is that the Labrador Duck's flesh was, in Alexander Wilson's understated phrase, "dry, and [partook] considerably of the nature of its food." That is, the bird tasted strongly of shellfish. One naturalist recalled seeing Labrador Ducks for sale at New York City's Fulton Market in the 1850s: ". . . Six fine males . . . hung in the market until spoiled for the want of a purchaser; they were not considered desirable for the table . . ." The Labrador Duck was, therefore, generally treated as poor folks' fowl.

The powerful flavor could be dealt with, however. In his 1819 history of Newfoundland, the Reverend Lewis Amadeus Anspach recommended carefully removing the skin to rid ducks of a fishy taste. "Priscilla Homespun" offered advice on dealing with this problem in her subtly titled, 1818 volume, *The Universal Receipt Book; Being a Compendious Repository of Practical Information in Cookery, Preserving, Pickling, Distilling, and All the Branches of Domestic Economy. To which is Added, Some Advice to Farmers.*

> One of the most effectual and simple expedients for destroying the fishy taste in water fowl, is by well washing their insides with vinegar and water, then, drying them thoroughly with a cloth and plentifully stuffing them with chopped sage, onions, eschalots, and sometimes even garlic, seasoned with finely powdered pepper and salt, and a little Cayenne pepper.

After roasting, basting and dredging, the duck was ready.

One 1860 cookbook noted that some people "with peculiar tastes" actually enjoyed a strong fishy tang in their fowl. And the native peoples of Canada even today consume wild birds whose flavor would deter many of European descent. Interestingly, biologist Glen Chilton, in his recent work on the Labrador Duck, unearthed two sources that described the bird as good eating; one eighteenth-century visitor to Labrador even called the duck "an excellent supper." Perhaps in that case, hunger made a fine sauce.

If the Labrador Duck usually was not popular to eat, what can we say of the relation between its decline and the astonishingly intense market-hunting of the nineteenth century? I think it not inconceivable that hunters sometimes killed the bird *because* it was such poor eating—a kind of mindless revenge on a creature deemed useless. As we have seen, nineteenth-century animal slaughters required little in the way of rational justifications on the part of the gunners. People shot anything and everything. So, too, with ducks, those well-flavored and those that were not. R. W. Shufeldt described the withering fusillades that ducks suffered on the East Coast:

> Those were the days [late 1860s] when the ducking-sloops used to come up from New York [to Long Island Sound] and kill ducks for the city markets. They were armed forward with a heavy swivel-gun that took a charge of a pound of double-B shot, and I have seen them sail down upon and fire into an acre or more of Old Squaws, killing and wounding some two hundred at a shot.

After the gun fired, men clambered into dinghies and rowed out among the ducks, using shotguns to finish them off. Then they gathered the dead. The result of such massacres was plenty of waterfowl to eat and declines in duck populations among all species.

Like shooting, "egging" disrupted and devastated many birds, including, probably, the Labrador Duck. (However, Glen Chilton points out that there is no written record of the collecting of a Labrador Duck egg.) "Eggers" could be local fishermen gathering eider or murre eggs as additional food. The eggers could be whole crews on schooners. These ships sought out rocky coastlines and islands in order to gather barrels of eggs (and feathers) for city markets in Canada and the United States. Audubon found the Labrador eggers a disgusting lot. He called them "assassins, who walk forward exultingly, and with their shouts mingling oaths and execrations. Look at them. See how they crush the chick within its shell, how they trample on every egg in their way with their huge and clumsy boots." Drunken, brawling men, the eggers destroyed what they didn't gather and shot any bird they wished to. Audubon noted how just one Halifax crew took 40,000 eggs in just two months in 1832, earning $800. Twenty boats sailed to egg that year in the Labrador region. By 1891, "this illegal and nefarious occupation," as author Alpheus Spring Packard called it, attracted only a few hardy souls, men ready to bribe anyone they needed to in order to avoid arrest, men ready to crush their cargos instead of being captured. But the practice continued because enforcement of wildlife laws in remote, barely populated stretches of Canada was—and is—difficult.

Two rather more insidious factors may have played roles in the decline of the Labrador Duck. Naturalist Outram Bangs suggested that expanded human settlement on the Atlantic seaboard may have affected the shellfish upon which the ducks fed. Increased sewage effluent discharged into coastal shallows might have reduced shellfish numbers. Researchers point out that "epizootics"—disease epidemics among animals—may have swept through the underwater prairies of long-leaved, flowing eelgrass, diminishing the duck's prey. "This habitat," biologist David Kirk writes, "is susceptible to periodic epizootic crashes which can have a drastic effect both on populations of some bird species and their migration routes ... Similar crashes in the very small Labrador Duck population could [have been] catastrophic ... especially if coupled with human predation."

And, apparently, the species was never populous to begin with, perhaps because its breeding habitat was restricted for some reason. Could increased ice packs, in the Little Ice Age, from the 1200s to the 1850s, have developed around rocky islets where the Labrador Duck bred, making the species vulnerable to mammalian predators crossing the ice? It is also intriguing to note that the last true Ice Age initiated, 3 million years ago, a devastation in the marine mollusks of the western Atlantic, according to one scientist, and as recently as 18,000 years ago, the Chesapeake Bay region was a boreal ecosystem. Could the effects of the more recent Little Ice Age have depressed supplies of the duck's preferred shellfish prey?

Despite its seemingly scanty population in the 1850s, the Labrador Duck could still be found on sale in eastern city markets. In the 1860s, most of the Labrador Ducks available were females or immatures—a sign that the valuable, attractive males were going directly to collectors and, perhaps, that the males may have wintered farther north. The number of juveniles in the markets was also a sign that most birds were not reaching maturity. Nineteenth-century hunter Nicolas Pike noted that, over many years, he shot no more than 12 Labrador Ducks, and only 3 or 4 of those were males. "... I have never met more than two or three [of this species] at a time, and mostly single birds," Pike said.

By 1870, the Labrador Duck had vanished from the market stalls.

Over the next few years, with not a single Labrador Duck sighted, ornithologists began to admit the possibility that the species was gone. "No one anticipated that they might become extinct," wrote one observer in 1891, "and if they have, the cause thereof is a problem most desirable to solve, as it was surely not through man's agency." Naturalist and collector George Boardman shared this incredulity; he wrote to ornithologist William Dutcher on October 29, 1890:

> The Labrador Duck I procured without much trouble and if I had any duplicates sent to me I did not save them any more than I should have saved duplicates of Scoters or Old Squaws ... Anything ... that I already had mounted generally went to the manure heap ... It seems very strange that such a bird should become extinct as it was a good flier.

Boardman also thought that because the Labrador Duck was a strong diver its extinction was all the more baffling. The presumption that a bird's skills of strong flight and diving could save it from extinction seems, even in Boardman's day, remarkably simple-minded. Lest we judge Boardman too harshly, however, we should note that he tried to coax Barnum's Museum in New York City to donate its pair of stuffed Labrador Ducks to the Smithsonian. A fire destroyed Barnum's before such a transfer could be effected.

In 1891, William Dutcher was finishing an article on the Labrador Duck that consisted mostly of a list of circumstances surrounding the origins of extant specimens of the species. Such a list can seem pedantic and dull to the general reader, but specimens of unusual birds hold much interest to collectors and museums, even today. The Labrador Duck is a rare bird even in death, with only about 60 valuable specimens around the world. How valuable? Enough so that Soviet forces leaving Dresden after the Second World War stole a museum's Labrador Duck as a war trophy! Years later, the Dresden curator traveled to Russia and had the duck returned.

As Dutcher worked on his specimen list, he decided to send copies of an illustration of the duck to two northbound expeditions, hoping that they might look for and inquire about the bird. Dutcher hoped, of course, for a rediscovery. A U.S. Navy-led trip to Greenland produced temporary excitement among the expedition's members. The native Greenlanders proclaimed they'd seen many such ducks. But upon further questioning, it became clear they had mistaken another species for the Labrador Duck.

On August 9, 1891, two men from a Bowdoin College Labrador expedition, which had moved 200 miles from the coast and up the Grand River, spotted a female duck with her brood. One of the men, a D. M. Cole, felt certain that the birds were Labrador Ducks. But Cole never saw the birds again nor found anyone who could confirm their identification and on that day he had no shotgun with which he could obtain specimens as proof.

Before the close of the nineteenth century, ornithologists were scanning the hunting magazines and ornithological journals for reports of the last shootings of Labrador Ducks. A shooting in 1871 and one in 1875 were cited by some authorities then, and by some now, as the last known instance of the species.

The American Ornithologists' Union, however, accepts an 1878 shooting as the last known occurrence of a Labrador Duck. The incident still raises questions, though, because it happened well inland from the duck's usual haunts and because the man who ended up with the specimen eventually lost it.

On December 14, 1878, an Elmira, New York, druggist named W. H.

*The illustration that William Dutcher provided two 1891 expeditions,
which he hoped might rediscover the Labrador Duck.*

Gregg sat down and quickly wrote a three-sentence note to the *American
Naturalist*. He reported an amazing incident: A Pied Duck had been shot
in Elmira just two days before, on December 12. Though Gregg did not
record the sex of the bird, he did use one of the then-current scientific
names for the Labrador Duck, so it was not another species. Gregg remarked that the bird had become "rare everywhere, and its occurrence so
far south in the interior gives special interest to the subject." He gave no
further details.

Fortunately, William Dutcher asked Gregg about the circumstances of
the Elmira shooting and the fate of the specimen:

> Dr. Gregg informs me that the duck in question was shot by a lad . . . in
> a broad expanse of lowlands called the Buttonwoods. These had been
> overflowed by the Chemung River, during a freshet. The duck had been

eaten before he heard of its capture: never saw or was able to procure any-thing but the head and a portion of the neck. These were preserved for some years. Recently while moving his collection to New York City he entrusted the packing of his specimens to another person, and as the head cannot now be found he suspects that it was thrown away with some moth infested birds as of no interest or value.

Had Gregg not published his small note in a respected scientific journal, it seems likely that the Elmira record would have been universally dis-missed and Dutcher might not have followed up on the particulars.

No notice of the shooting appeared in the Elmira press, though many articles covered a terrible storm that must have forced the unfortunate bird inland. The temperature on Thursday, December 12, 1878, hovered in the low to middle 30s, and the waters roared. "The high water," wrote the editors at the *Elmira Daily Advertiser*, "has done untold damage in nearly every part of the country ... The storm extends to all the Eastern, Middle and Southern states." In Elmira's Buttonwoods, an Irish settle-ment on the south side of the Chemung River, the floodwaters "were rampant, and went rushing and tearing like mad ..." the newspaper re-ported. No dike had yet been built to protect the Buttonwoods, which sat squarely in a floodplain crossed by two big streams. (The name Button-woods has two meanings; it can refer to plane trees or sycamores, but also to the shrubby buttonbush, which thrives in wet soils and produces seeds favored by ducks and shorebirds.)

On the north side of the river, the cellars of many local businesses filled with water. W. H. Gregg's pharmacy at 333 E. Water Street must have been flooded, as was, perhaps his home on 153 Madison Avenue. While knowing this context may not compensate for the loss of the spec-imen, the fact that Gregg made the time to write a note to the *American Naturalist* in the midst of chaos and clean-up underscores just how sig-nificant he found the shooting of the Labrador Duck.

Today, what used to be called the Buttonwoods includes Brand Park,

Riverside School, Miller Pond, Coldbrook Creek and a neighborhood whose streets bear names such as Maple, Catherine, Liberty and Robinson. A levee now rises between the neighborhood and the Chemung River. From the grassy dike, one can look north at the river and see the buildings lining Water Street. Where Gregg's office stood there now rises the four-story Elmira Savings Bank. A car dealership and a convenience store occupy the site of his former home on Madison. Big cottonwoods have grown up on the slope of the levee leading down to the river. In the area immediately south of the dike, maples grow that could be a century old, and just shy of where Catherine Street bends south and becomes Robinson, several big sycamores stand in stately fashion. Some of them must have been saplings on the day the last Labrador Duck was shot.

On a summer day, the river fractures reflected sunlight. Herring Gulls swoop above the Sly Street Bridge, where the ferry used to cross, and a long shoal of sand and river stones creates dark ripples in the current. A man picks out cans from trash containers and pauses to inspect his collection next to a copper marker set in granite. The sign notes a "Memorial Tree" dedicated by the Elmira Rotary Club to Lieutenant Harry B. Bentley, killed in combat in World War I France. There is no tree anywhere near the sign. Next to a sawed-off stump of a metal pole, there is another marker. This one proclaims, "Presented by Head Camp Woodmen of the World."

Here, by a river the Algonquins named Chemung—their word for "place of the big horn," because the tribe found Woolly Mammoth tusks along the banks—one can ponder the loss of an elusive, handsome duck. One can look south from the levee into Brand Park and the surrounding yards and imagine a young boy toting a black-and-white duck back home to his hungry family who, quite understandably, would not have known or even cared about the fate of this species.

The Labrador Duck was the first strictly North American bird to become extinct in historical times. Its loss preceded by many years the careful management of waterfowl in North America. Today, through a

Part of the area formerly known as the Buttonwoods in Elmira, New York, where the last known Labrador Duck was shot on December 12, 1878.

combination of education, government regulation, wetlands protection and financial support from hunters, birders and conservation organizations, waterfowl on this continent have reached all-time record high population levels. In fall 1997, the U.S. Fish and Wildlife Service tallied 92 million ducks of all species. That number declined to 84 million in 1998, but surged in 1999 to 105 million ducks. Concern remains, though, for Northern Pintails, sea ducks and for scaups. Continued attacks on wetlands by developers and corporate agriculture should keep us from complacency, as should the story of the Labrador Duck.

> Interred within the few skins, mounts, and bones [of the Labrador Duck] that are scattered throughout the world's museums like the deteriorating leaves of a now-dead oak are the genes and chromosomes that represented the species' strategy for survival in a hostile world.
>
> —Paul Johnsgard, *Waterfowl of North America*

The images from *Jurassic Park* and thoughts of cloning the dead back to life troubled me for a long time. Despite having read articles about Dolly, the cloned sheep born in 1996, I didn't really understand anything about cloning *living* things. And to learn if we could resurrect *extinct* creatures—such as Labrador Ducks—I needed to review just what DNA is, understand the current status of cloning technology and learn about research that uses extinct DNA.

In complex lifeforms, DNA consists of a precisely ordered set of molecules. It resides in cell nuclei as tight bundles of chromosomes. (To get an idea of just how tightly packed DNA is, consider that, in humans, just one cell contains six feet of DNA; if all the DNA in an adult human body were stretched out, it would reach to the sun and back.) The molecules in DNA determine the sequence of amino acids in proteins. The nucleotides that hold the famous double helix of DNA together—adenine, guanine, cytosine and thymine—vary in sequence from individual to individual. DNA is an individual creature's genetic blueprint. Writer and scientist Chet Raymo puts it more poetically. DNA "is a pulsing, undulating farrago of threads, feathers, knobs, and whiskers, a microscopic lace maker frenetically making a lace called life."

It used to be that if one wanted to create an individual member of an animal species, two sources of DNA were needed—a male and a female and some quiet (or not-so-quiet) time. From the union of two sets of DNA comes a third, unique set of DNA. A new individual. Offspring.

While this is still, thankfully, the preferred way of doing things, there are options. If you happen to be a mouse in Japan or a sheep in Scotland, you might have arrived on this mortal coil in an entirely different way. You might be a clone—offspring, yes, but from a single parent.

How is this possible? The procedure that created Dolly, the world's first mammal cloned from an adult cell, works like this: A researcher takes intact and complete cells from a donor sheep and slowly reduces the nutrients those cells need for sustenance. In this way, the cells become quiescent. They pass into a stasis of nonduplication. Then a micropipette is used "to suck out the chromosomes," writes Dolly creator, Ian Wilmut. Thus, an unfertilized egg cell of a recipient sheep is emptied of its DNA. The DNA from a quiescent donor cell is next injected into the empty egg cell. Controlled electrical pulses stimulate the concoction to begin growing—to, as it were, wake up. Researchers implant the newly filled egg into the uterus of a third sheep, which carries it to term. This procedure means that we can now "reset" DNA in adult, nonembryonic cells to create an entire animal, the exact genetic duplicate of the donor.

Even in the face of many technical difficulties associated with cloning live creatures in this manner—problems such as unforeseen cellular mutations, apparent quickened aging of clones and a high degree of mortality for cloned cells and their "offspring"—scientists continue to pursue and refine this perceived betterment on millions of years of evolution.

Despite the attention given to the possibility of cloning humans (which deserves glaring scrutiny), the current cloning effort primarily focuses on duplicating genetic material of animals and, as in the case of Dolly, copying whole individual animals. But there is an additional aspect to this work. Companies are especially interested in creating "transgenic" creatures, animals that have incorporated into their bodies little bits of DNA from a different species. This allows researchers to create, for example, a sheep that produces a new kind of milk—milk containing chemical compounds that can be used to prevent or treat certain diseases. The compounds are "manufactured" by the DNA from the other species that now resides in the sheep. In short, the sheep produces its milk as always, but the imported DNA injects into the milk chemicals that the scientists—and chief executive officers—hope will be medically and financially viable. To make more such sheep, you can clone them using the Dolly method. Eventually, you'll have a profitable assembly line of perfectly identical, cloned transgenic critters dispensing their pharmaceutical milk. On such dreams are fortunes made.

Research using DNA cloned from *extinct* species also moves forward, though with far less public fanfare. Such work began in 1984, when Allan C. Wilson led a team that cloned portions of extinct quagga DNA that he had gathered from an old museum skin. (A horselike animal of southern Africa, the quagga went extinct in the nineteenth century.) Wilson laboriously inserted into the bacteria what quagga DNA he had isolated, and the bacteria copied the quagga material. This process was enormously time-consuming, but copying today is much easier thanks to a now-commonplace procedure called polymerase chain reaction, which quickly duplicates DNA.

So what's to keep us from booking tickets to the grand opening of Jurassic Park? Apart from zoning considerations (and dinosaur habitat needs), the scientific difficulties of getting started on such a cloning enterprise are staggering. Cloning a whole life-form, a living, breathing, used-to-be extinct creature from extinct DNA is as difficult as the image of its success is compelling. To clone a stegosaurus, or any dinosaur, you have to locate relatively intact dino DNA. Even if you manage to find an insect trapped in amber, you can't be sure the insect's blood contains the DNA of a "terrible lizard." Getting it out risks contamination from DNA in the present environment. And how do you know what DNA it is that you've found? Your dino dreams run into the reality of limitations pretty quickly.

But what about *recently* extinct birds? After all, the birds of which I write all existed within the past 200 years, much closer in time than triceratops. Researchers have sequenced DNA from extinct birds, but the purpose in mind was phylogeny (how an extinct creature relates to other species in time), not resurrection. Oxford University's Alan Cooper once sequenced short sections of DNA he recovered from Dodos held in museum collections. Work of this sort enables study of how the famous, flightless Dodo compares with related birds, such as pigeons, through the course of their evolutionary changes. Illuminating the life of an extinct species may help us know what is required to prevent other extinctions today (apart from calling a halt to destruction of habitats).

Of course, sequencing and examining bits of extinct bird DNA is one thing. Using that as the basis for cloning an entire bird from those bits is quite another thing. "It's not possible now," says Rob DeSalle, an associate curator at the American Museum of Natural History and an expert in extinct DNA. "That doesn't mean we couldn't do it eventually." (DeSalle also notes that if gametes from a Passenger Pigeon had been frozen before its death, there would be a reasonably good chance of bringing the bird back to life.)

What are the technical challenges that prevent us from cloning an extinct bird back to life? For one, all dead things—including the many museum specimens of extinct North American birds—"lose" their genetic material over time. That's why Wilson got only bits of quagga DNA, and Cooper, bits of Dodo DNA. Researchers disagree about the estimated rates of DNA degradation. One researcher has said that DNA should be, in general, irretrievable after only 50,000 years, if not far fewer. Others contend that at least scraps of DNA can remain intact for tens of millions of years. What is certain is that at least some DNA sequences will be missing from any genetic material extracted from an extinct creature. (Even when specimens are preserved in formaldehyde, they suffer the same loss of DNA, since formaldehyde, like water, actually accelerates breakdown.) DeSalle calls this vexing matter of missing sequences "the Humpty-Dumpty problem." To retrieve bird DNA that may be less broken up as well as less likely to be contaminated from the present environment, researchers must drill into bone or scrape off dried muscle tissue rather than retrieve it from the surface of the specimen.

If you want to clone DNA in order to coax out a resurrected bird, you will first need to figure out how to replace any missing sequences with portions from closely related species. This would require, among other things, the sequencing of genetic codes of several bird species—which is difficult work—in order to understand what is missing. If you were attempting to clone a Labrador Duck, but didn't have the DNA coded for its beak, you could try to splice in similarly coded DNA from a Steller's Eider. This would be transgenic research at the outer limits.

Some scientists predict that transgenic work with extinct DNA will become quite routine. These efforts would focus on supplementing already-living species with recovered fragments of extinct DNA. In such projects, ancient DNA from a wide range of insects and plants might be put to use today. For example, a present-day crop could become more resistant to disease or drought once it is supplemented with ancient DNA coded for those traits.

Given the pace of progress in cloning and the anticipation of even more advanced application—despite the difficulties already discussed—I ask Rob DeSalle the question uppermost in my mind: "So, it's not laughable, is it, to imagine that we might be able to clone an extinct bird back to life?"

"Oh no," he says.

But when asked the same question, Oxford's Alan Cooper couldn't disagree more. He sets forth an articulate, impassioned dissertation of doubt, and zeros in on the two crucial problems facing the potential cloner of extinct birds. Anyone trying to recreate an entire extinct bird, Cooper says, will *never* be able to solve the Humpty-Dumpty problem. Using a different metaphor, he says, "It's the world's biggest jigsaw puzzle." If you have a 30-million-base pair genome, how do you finish the puzzle with only, say, 400 pieces?

Cooper notes that whatever Labrador Duck DNA you have—or any such extinct DNA—would then need to be reassembled into chromosomes, requiring proteins encoded within the DNA itself, requiring other proteins to make those proteins and allow cell division to occur. Some of these proteins would be *specific* to the Labrador Duck alone. Trying to splice in DNA from a related species—using the Steller's Eider DNA coded for "beak" to fill in the gap with your Labrador Duck sequence— is not as simple as the verb "splice" connotes. It's like taking differently sized and shaped parts from another puzzle to complete the one you have. According to Cooper, chances are they won't fit. It's the old round-peg-in-a-square-hole problem. "DNA and proteins interact in a way that is specific to that [species or genus] or perhaps very close relatives," Cooper says. "Furthermore, a large part of the genome is noncoding . . . [and] often evolves relatively quickly, so that even closely related taxa would be unsuitable." And there's the not-so-small problem of getting a bird of one species to carry to term an egg cell filled with the DNA of another species.

Alan Cooper can't foresee any future advancements that would change his judgment that the prospect of cloning an extinct species back to life is, in his words, "complete rubbish."

DeSalle and David Lindley, a respected science writer and editor, pointedly remind us, however, that "the sorts of problems that remain perplexing—how to identify and 'mend' unknown genes, how to persuade an egg cell to take up a foreign genome, how to manipulate the genetic characteristics of a growing embryo—are those that scientists are working on right now." These seemingly insurmountable difficulties might be solved in the not-so-distant future.

For the time being, cloning does remain a viable option for conservationists working only with still-living species and those recently extinct whose remains or fluids have been frozen. After Dolly's birth, a Utah State researcher was considering using the Dolly method to clone an endangered relative of sheep in China. Biologists already anticipate that zoos may find cloning procedures an additional tool in the preservation of plant and animal DNA.

As regards to extinct species—the ones already lost, whose shattered DNA is available only in museums—Cooper's skepticism is not unjustified. The odds remain fantastically remote that we will be able to clone an extinct bird back to life or that we would even choose to expend the resources to try—resources better used to protect, conserve and restore habitats and species still with us. Says Cooper, "It is actually dangerous to encourage people to think that this may be possible because it down-grades the seriousness of extinction . . . Far better to try and save what is left in the world than devote time and money to science fiction pipe-dreams."

Yet DeSalle and Lindley's comments—indeed, the entire history of Western science—remind us that we are nothing if not dogged in our cleverness. We imagine. We pursue. Some scientists, for example, despite the many technical challenges involved, are executing a plan to retrieve and resurrect ancient DNA not simply for study, but to recreate a living member of an extinct species. Japanese physiologist Kazufumi Goto and geneticist Akira Iritani hope to locate a frozen Woolly Mammoth in Siberia. Should the animal be well preserved, they will try to extract frozen sperm to use in an effort to birth a Woolly Mammoth through in-vitro fertilization and cross-breeding with modern elephants, itself a tricky proposition. If the researchers can locate any DNA that is more or less intact, they also could attempt to recreate a Woolly Mammoth through cloning. (I, for one, am struck by the absurdity of attempting to bring back a Woolly Mammoth at the apparent dawn of a much-warmer era on our greenhouse planet.) In 1990, Goto used cow sperm that was lifeless to fertilize a living cow; a calf resulted. So even if we dismiss this dream as far-fetched and misguided, Goto is a serious scientist.

Considering all of the controversies over cloning, I keep asking my-self about the birds. What if DeSalle's optimism is realized? What if, someday, we could return a Labrador Duck to the sand shoals of Long Is-land or the rocky islets of the Gulf of St. Lawrence? Of course I'd be awed by a world in which birds that humans had destroyed could be restored, wholly, in places that we resuscitate back to a semblance of a healthful biodiversity. Such a restoration could be viewed, after all, as reparation for our ancestors' blind voracity, the spewing swivel guns of ducking-sloops. After such an improbable healing, though, would we renounce the means that got us there? Will assuming genetic control over the fates of species feed our enormous selfishness in relation to all these lives who have been here longer? When has modern culture truly renounced a scientific power it has gained?

In the short term, we do control the fates of species and habitats. But that control should be woven with humility and respect, not the arrogance of assumed mastery. I suppose it's far more likely that even if we could find the means to resurrect Labrador Ducks we would restore them only to climate-controlled zoos and not to an ocean now afflicted with pollution and overfishing. Thinking about cloning cuts directly to the question of how we view ourselves in the world around us. Do we act like gods or do we live as acolytes?

Imagining vanished lives helps answer that question. Knowing whatever we can about these vanished birds restores them, after all, to a habitat we still can save: our moral imagination. It is in that place that we glimpse—and only glimpse, through fact and dream—the most mysterious of these birds, the one about which we know so little with certainty: the Labrador Duck, cipher of the strangest sea.

In a Northern Gulf: Journey to Bird Rock

IN A COOL MORNING BREEZE, ELIZABETH AND I STOOD BY THE EDGE OF THE Grosse-Île harbor, looking at the boat we would soon board. It appeared perfectly sturdy, even spiffy, but very, very open. No roof. No cabin. This was not a comfy craft for Chablis-sipping tourists putzing around some pleasant, protected inlet. This 24-foot Zodiac seemed more like a surfboard with sides. A surfboard on steroids. Or perhaps a white-water raft with glandular problems. It had a fiberglass bottom, but no seats (except for one behind the captain's tiny steering console). At the stern, there was a kind of aluminum roll bar. Atop the boat's inflated, black rubber sides, all along on a stiff red brim, was braided a white rope like bunting.

"Look," I said to Elizabeth. "Ropes. They're for hanging on."

Low-slung and fast, Zodiac boats look like those sleek assault crafts that haul commandos to hidden beaches for whatever it is commandos do on hidden beaches. It's very Jacques Cousteau, I thought, very *not* Marlin Perkins. The boat's proud and savvy owners, Gaston Arseneau and Nadine Blacquière, busied themselves. They had to off-load the vessel from its trailer and slip the boat into the water.

I glanced at my wife to see what apprehension her face might betray. After all, we were going to ride this open boat—shades of Stephen Crane!—some 20 miles from this island harbor, out across the Gulf of St. Lawrence, then back again. Our destination was a hulking, sheer-cliffed island and three smaller nearby islets called Bird Rocks, one of the most important seabird nesting colonies in the world and one of a handful of known breeding sites for the extinct Great Auk.

This was our fourth day on the Îles de la Madeleine, a 40-mile-long chain of tombolo-connected islands north of Prince Edward Island. Part of a massive shoal breaching the Gulf's cold waters, the Îles de la Madeleine rise up on subterranean salt domes. The islands make up a portion of the northern end of the Appalachian Mountains. Appropriate to a place where many people still earn their livelihoods by harvesting the fruits de mer, the islands look, on a map, like a graceful fishhook. The archipelago belongs to Canada or, as most of the 14,000 residents would say, speaking French, they belong to Quebec.

After supporting generations of hunting-and-fishing treks by Micmac Indians, this remote place became a refuge for Acadian settlers in the 1750s. Political and economic control passed through many hands, including those of Isaac Coffin. A heartless autocrat, Coffin ruled the islands in the late eighteenth century as a king ruled serfs. Not until 1895 did the Madelinots gain the right to buy their own land. For years, the relative

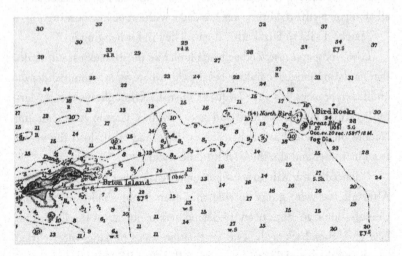

*Detail from an old chart, showing "Great Bird," the main island,
and nearby "North Bird," or Gull Rocks.*

The main island of Bird Rock, in the Gulf of St. Lawrence, covered with nesting Northern Gannets and other birds. Though the appearance of the sea-beaten cliffs has changed over the centuries, the lowest ledges visible in the photo are of the type that would have been accessible to Great Auks.

isolation of the islands worked against achieving equitable status even within federal and provincial democracy. The Îles de la Madeleine didn't become a full-fledged county until 1948.

Today, mainland Quebecois flock to the scenic islands each summer for vacations, and the local economy counts tourism as critically important. Winter vacationers visit the pack ice where seal-watching has mostly taken the place of seal hunting. (For two recent winters there has been no pack ice, which more than one islander matter-of-factly attributes to global warming.)

With steep sandstone cliffs, ocean-carved grottoes, grassy demoiselle hills, wide beaches and patches of wind-stunted conifer, elfinwood and alder thickets, the seven main Madeleine islands are stunningly beautiful. Georgia O'Keeffe would have loved the plunge of these red cliffs into green water, the guano-streaked rocks with their myriad folds and tucks and the black holes of sea caves. Treeless hills rise, typically, within a mile of shore, creating a sense of insistent verticality as well as a vast horizon. There are quaint harbors and brightly colored houses, low rolling plains and duck-beckoning lagoons, stretches of fragile dunes and headlands of gray-green sandstone all packed into 150 square miles of surface area. Yet these islands feel spacious.

Elizabeth and I had arrived in mid-June to avoid the July-August tourist peak and to coincide with the best time to observe nesting seabirds at Bird Rocks, or Rochers aux Oiseaux. We spent days hiking, visiting museums and researching the human and avian history of this intriguing place. We were also waiting for the waters to calm. Each evening we'd call Gaston and Nadine to ask if the next morning we'd be able to head out to Bird Rocks. I had begun to worry we'd run out of time and have to return to Kansas without even attempting the trip. "We don't have to *land* on the main Bird Rock," I told Elizabeth one night, which she relayed in French to Gaston. (Elizabeth used her college-major French to good effect while I stood dumbly by.) "Tell him I just want to see the place."

A couple of hours before we arrived at the harbor, Gaston had pulled up to the chalet we were renting and gave me the news. In careful English he said, "The sea is, uh . . ." He smiled and shook his hand.

"Rough?" I said, as the wind whipped a dish towel we had secured to the clothesline.

He smiled. "We'll try. If it's too bad, we'll come back, okay? Bring some clothes!" He pretended to pull on a jacket and shiver.

After we arrived at the harbor at Grosse-Île, which sits at the northern end of the main archipelago, I studied the water. It looked calm enough. The wind, though, the wind. Using my best face-in-a-Kansas-cold-front powers of estimation, I figured the gusts were reaching about 15 or 20 miles per hour. While I checked over our gear, Elizabeth chatted with Pascal Arseneau, a stocky, handsome, thirtysomething islander who had given us a ride here. Pascal, no relation to Gaston, is a photographer and one of the officials of the local tourism association. Like many native islanders, he had never been to Bird Rocks and was anxious to see them, even though he also knew a journey there was no trivial matter. As a Canadian Coast Guard crewman had told me earlier, in a thick accent, "It is, you know, a very wild place."

Elizabeth and I jogged up to a building to use the restroom; then, as we were leaving, the fisherman who had let us in cast a dubious look our way. Like nearly all the residents at this end of the islands, he was an Anglophone, one of about 750 English-speakers on what *they* call the Magdalen Islands. With a Scots-Irish lilt in his accent, he warned, "The wind's against the tide today." On four or five hours of sleep, I had to *think* . . . what does that mean . . . the wind's against the tide today. High swells. A strong head wind on the journey back from Bird Rock, if, in fact, we made it there.

"Whereyagoing?" the fisherman asked.

"Bird Rocks."

"In dat ting?" He cocked his head to indicate the harbor and Zodiac boat.

Elizabeth and I nodded.

"Whostakingya?"

"Gaston Arseneau," Elizabeth answered.

The fisherman's face brightened. He nodded. "Gaston, yup, I know 'm. Good man."

A former barkeep and fisherman, Gaston owns and operates Excursions en Mer with the lively and bright-eyed Nadine, his partner in life and business. They take tourists on cliff-viewing trips at sunset, hikers to nearby Entry Island, and amateur anglers out to catch what they can in these overfished waters. Gaston is tall and muscular: This is a man who "ice-canoes" for fun, who, with his Roman nose, dark sunglasses and thinning gray hair, looks like a Quebecois Dennis Hopper in a life vest.

Nadine loaned us expensive rain gear to supplement the jackets, long johns and wool sweaters we'd brought with us. She inflated the pads on which we'd sit, and motioned us into the svelte Zodiac. It was time to go.

Because the trip from Grosse-Île to Rochers aux Oiseaux takes about three hours one-way in one of Nadine and Gaston's slow boats, they hoped this faster craft would attract extremophile birders and photographers. That was when I learned the trip to Bird Rocks was a research expedition for them as well: They had received the boat only a few days before. Gaston stoically steered us out of the harbor and northeast across waters rich with shoals and shipwrecks (perhaps 1,000 of the latter).

As the boat sped outward, beneath clear skies, I put my binoculars away. The seas were choppy and any view through the lenses felt dizzying. Pascal, Nadine, Elizabeth and I small-talked, then fell silent as the engine droned on. I spread my legs out on the floor of the boat and leaned against the side, watching the ocean and some distant gulls. For the first hour of my adventure, I saw few birds. I thought of only one.

June 16, 1534. The explorer Jacques Cartier arrived at Rochers aux Oiseaux. With Cartier began the written record of the Great Auk in the Gulf of St. Lawrence. He named this flightless bird of the open ocean the "Apponatz," which apparently derives from a word of the long-vanished Beothuk Indians of Newfoundland. Cartier found the islands to be "as full of birds as any field or meadow is of grass." He wrote that "we went down to the lowest part of the least Island, where we killed above a thousand of those Godetz [Razorbills] and Apponatz [Great Auks]. We put into our boats so many of them as we pleased, for in less than one hour, we might have filled thirty such boats of them . . ."

A bird the size of a large goose, the Great Auk looked somewhat like a fat torpedo. Its plumage was white below and black above, except for a white patch in front of each hazel- or chestnut-colored eye. Its huge grooved beak appeared both unwieldy and menacing. Its wings seemed absurdly small, only a few inches long.

Sometime in the course of its evolutionary history, the Great Auk gave up flight to become a specialist: a diving bird able to stay submerged for perhaps as long as 15 minutes, capable of reaching depths of 250 feet or more. It's even been claimed that the Great Auk could dive perhaps as far as a kilometer. Whatever depths it reached, it reached them swiftly. A Scottish professor who kept a captive Great Auk—it eventually escaped—allowed the bird to roam in a harbor while attached to a tether. He wrote that "even in this state of restraint it performed the motions of diving and swimming under water with a rapidity that set all pursuit from a boat at defiance."

Great Auks would hurtle through the cold North Atlantic water, catching lumpsuckers and shorthorn sculpins and perhaps cod, sand lance and other fishes. The bird's complete diet is not known with certainty. It is clear, however, that the Great Auk preferred to feed in shoaling waters. In the relatively shallow waters of the outer continental shelves, such as the Grand Banks of Newfoundland and Labrador cold and warm ocean currents mix and create a lush zone of phytoplankton. A marine ecosystem there should swarm with life: Zooplankton eat phytoplankton, krill eat zooplankton, fish and whales eat krill, and onward the watery web extends. Cod once thronged the Grand Banks so abundantly that sailors had difficulty rowing dinghies through their dense schools. Seabirds—shearwaters, gannets, alcids and more—gleaned fish in dizzying arrays. In such places, Great Auks would float and bob with the swells, like grotesquely oversized ducks, their necks pulled in but their heads lifted up as if on constant, polite surveillance. "When they are seen a great way off, in flocks, they look very much like a number of children, dressed in black, with white aprons on," stated the 1830 *Child's Book of Nature*.

The Great Auk was found in an arc that extended from Gibraltar up the European coast to the North Atlantic, west to Iceland, Greenland, then south to Newfoundland, the Gulf of St. Lawrence, and down the American coast as far as Florida. (Audubon reported that the species had once been commonly seen in winter off the Massachusetts coast). The southerly locales of Italy, France and Gibraltar in Europe and Florida in North America probably marked the limits of the species' pelagic wintering range.

The Great Auk's legs were set far back on its body, an adaptation for powerful swimming and diving. This design strength at sea put the bird at a disadvantage on land. In late spring and early summer, the Great Auk had to come to land to breed. It could do so only on rocky islands with gently sloping shelving that allowed the bird to ride the waves in, hop upright and awkwardly waddle toward points above the watermark.

Given its restrictive need, the Great Auk may have had no more than about 20 breeding colonies. (For a time, many mistankenly believed the Great Auk was a bird of the Arctic.) Biologists can confirm only eight islands, in fact, where we know Great Auks gathered to breed: Bird Rocks in the Gulf of St. Lawrence; Penguin Island and Funk Island, both off the Newfoundland coast; Geirfuglasker and Eldey Island near Iceland; the Faeroe Islands, between Iceland and the British Isles; St. Kilda off Scotland; and Papa Westray in the Orkneys. Some records written in the seventeenth century suggest that the birds may have bred as far south as Cape Cod. In the early 1600s, at Cape Cod, explorer Bartholomew Gosnold saw "penguins," which was then a common term for Great Auks. John Josselyn, traveling in New England in about 1670, spoke of a flightless bird called a "Wobble," a name that could only have been applied to a Great Auk on land.

In mid-to-late June, the female of a mated pair would lay one egg as large as a man's hand. The egg, buff-yellow to white, and marked with dark splotches and streaks, was simply laid on bare rock. Its pyriform (or pear) shape may have protected the egg if it began to roll. Many authors have stated that this shape ensured the rolling egg would describe a circle, which presumably reduced accidental loss. But biologist Tim Birkhead points out that although the pyriform eggs of other species do indeed roll in a circle, the circle those eggs describe is wider than the ledge on which they are placed. At least everyone agrees that Great Auks would stand in an upright position to incubate the eggs.

The chick would hatch out after six weeks, then venture to sea with its parents after only five to nine days at the colony. This rapid departure may have spared parents numerous energy-wasting trips to gather food to bring back to the colony. At sea, the young probably fed on plankton but may have taken regurgitated fish from the parents as well. No one ever captured—and preserved—a chick specimen, but the one naturalist who claimed to have seen chicks at sea said they were covered in a plumage of gray down. Nicholas Denys, the seventeenth-century governor of New France, claimed that the chicks rode their parents' backs, though this is a behavior never actually observed in living species related to the Great Auk.

Breeding probably did not take place until a bird was several years old, and ornithologists estimate that Great Auks lived to the age of 20 to 25. They also believe that males and females paired for life. These birds felt strongly, it seems, their fidelities to place and to each other.

Just how many Great Auks once plied the ocean is impossible to say. Some experts argue that millions of them lived at the time of European discovery of the New World; others contend that the species could not have been so populous as that because increased ice packs during the Little Ice Age (1200s–1850s) restricted the bird's breeding range. Breeding sites accessible by pack ice or floes would have been vulnerable to predation by south-ranging polar bears and other mammals.

We do know the species was familiar to, and hunted by, humans for thousands of years. Newfoundland's extinct Beothuk Indians used the dried yolk of Great Auk eggs to make a pudding. Archaeologists have discovered Great Auk bones in middens from Florida to Norway. At Gibraltar and at Jersey, in the English Channel, bones that date from 70,000 to 90,000 years ago have been uncovered. An exceptionally artifact-rich cemetery site at Port au Choix in Newfoundland yielded insights into the Maritime Archaic people's interactions with a range of bird species. This 3,000-to-4,000-year-old site contained the bodies of 53 humans—and more Great Auk remains than of any other bird. One human had been buried with 200 Great Auk beaks, part of a ceremonial shroud or blanket. In a cave at Grotte Cosquer, France, someone painted Great Auks 20,000 years ago. The figures are still visible.

While the Beothuk Indians spoke of Apponath, or Apponatz, the Inuit called the Great Auk Isarukitsck, which meant Little Wing. In Iceland, the bird was "Geirfugl." Spearbill. The Dutch called it "Tossefugl," or "Stupid Bird," for reasons that will soon come clear. Some called it the Garefowl. Many mariners spoke of penguins, a name eventually transferred to the unrelated flightless birds of the southern hemisphere. (The Great Auk's scientific designation still bears this old lineage: *Pinguinus impennis*.)

In *The English Pilot*, an eighteenth-century book of navigation aids, captains and crews were advised to watch for vast numbers of seabirds as their ships approached the New World. The presence of such birds meant the Grand Banks were near. Of the birds to watch for, "none are so much to be minded as the *Pengwins*, for these never go without [beyond] the bank as others do."

Sailors from the Old World found these birds to be more than markers of place. Great Auks were food. Before venturing to Rochers aux

Great Auks (drawn rather crudely) for The English Pilot, *a navigation book for mariners. This is from the 1780 reprint of the 1767 edition.*

Oiseaux in the Gulf of St. Lawrence, Cartier's expedition had discovered the birds breeding on an island off the Newfoundland coast. The place came to be known as Funk Island, so named because "funk" denoted stench, and hundreds of thousands of nesting seabirds will produce just that.

Funk has another meaning: panic. An emotion that the Great Auks must have felt *somehow*, though their behavior seemed a kind of stunned complicity. "These Penguins . . . multiply so infinitely upon a certain flat island that men drive them from hence upon a board, into their boat by the hundreds at a time, as if God had made the innocency of so poor a creature to become such an admirable instrument for the sustenation of man." So wrote Richard Whitbourne, a seventeenth-century chronicler of Newfoundland. Whitbourne, who also reports in his writings the sighting of a mermaid, should nonetheless be trusted on this particular account. He was present when Great Auks were herded from Funk Island onto a vessel and then beaten to death and eaten or stowed for later sustenation. Although biologist Tim Birkhead questions whether the birds would have allowed themselves to be herded on planks or taut sails, historical sources insist this was, in fact, the practice.

The massive colony at Funk Island probably supported some 200,000 Great Auks during their brief annual landfall for breeding, according to biologists William Montevecchi and Leslie Tuck. These birds, wherever they bred, supported mariners, who, short on supplies, found the "penguins" deliciously fat and tasty. Though startled by noises, Great Auks usually did not react with fear to what they *saw*, so quietly rounding up or knocking down the birds proved easy enough. A person might risk a bite from the sharp bill, but, otherwise, the birds hardly put up a fight. Men also gathered their eggs, which were big and healthful. Because the birds did not defend them, a crew could gather thousands of eggs in just a single day.

These facts formed the basis of the opinion that Great Auks were "stupid," an assessment we now know misses an aspect of evolution: The

*A "harvest" of Great Auks along the Newfoundland coast.
This engraving, by F. W. Keyl and E. Evans, appears in* Links in
the Chain, *an 1880 book by G. Kearley.*

birds had evolved without the constant presence of terrestrial predators on their breeding sites, a fact that explains their tractable behavior when humans attacked. Further, Great Auks had not evolved a breeding strategy that allowed them to sustain themselves in the face of relentless persecution. Being site-dependent, they could not disperse to other nesting grounds. In their element—diving in ocean waters—Great Auks demonstrated grace and puissance. No Cartier or Whitbourne could do what a Great Auk could do, and our submarine craft will never be as agile as that departed bird. If any species earned the label "stupid," insofar as humans and Great Auks were concerned, it was ours—in particular, the Europeans who settled the New World. They *could* have harvested Great Auks at levels that would have allowed the birds to continue living indefinitely.

That, in fact, was what native peoples in the North Atlantic and coastal northeastern Canada did for centuries. They hunted the Great Auk and they also revered it. They were grateful for what it provided. They did not overexploit the species. But once fishing vessels from England, France, Portugal and Spain began in quantity to reap the riches of New World fisheries, Great Auks and other seabirds were fair game for excessive killing. As early as 1578, some 350 European fishing vessels worked the Grand Banks. That is a fair indication of the number of crews gleaning wild food in the form of cod, mackerel—and Great Auk.

In northern European waters, Great Auks were also hunted. One Icelander expert in capturing them, reported that "the wings are kept close to the sides when the bird is at rest, but a little out (so that light shows under it) when it begins to run." That is, run from humans.

On the continuous gelatinous crease and crest of water, on the dip and swell of waves, Gaston steered his jaunty boat. I found myself distracted by the increasing numbers of birds. More and more the air filled with the heavy sleekness of Northern Gannets, white birds with black wing tips, a golden head and a blue-gray bill. Called Fou-de-Bassan in French, "Fools of the Basin," they flew low and high, in all directions. We looked up to watch them soar and turn and, in doing so, we kept our mouths tightly shut. Flecks of guano decorated the boat's black rubber sides. A gannet will wheel quickly overwing and plunge into the water from as high as 100 feet in pursuit of fish. These birds looked to me like supersonic bombers, inappropriate as this simile may be.

I held on to the Zodiac's rope and saw more birds appear in the air. We were nearing Bird Rocks, heading first toward what is sometimes still called Great Bird, the main island with its 100-foot cliff of red and gray-green sandstone and a surface area of some six acres. Bird Rock—singular, meaning this main island—is home to an automated lighthouse and thousands upon thousands of nesting Northern Gannets, Atlantic Puffins, Leach's Storm-Petrels, Thick-billed Murres, Common Murres and Black-legged Kittiwakes.

With the wind behind us, the air felt warm and pleasant, even if the 10- to 12-foot swells kept Elizabeth and me feeling tentative, almost woozy. "Big water," Elizabeth said, and I nodded, wide-eyed.

Three times the new engine cut out, putting us adrift with no reassuring sense of forward motion. The wind would catch up, chilling us and wafting gasoline fumes across our tiny craft. Three times Gaston turned his back, pulled out tools, tinkered with the pump and restarted the engine. This adventure was already worth whatever mechanical and physiological setbacks we had encountered, though I didn't relish the prospect of breaking out the oars for a long row home. I felt thankful for radios. A Common Loon raced above us like a high-speed blessing: We'd get to where we needed to be.

Again and again, the prow of the Zodiac lifted high like a stalling airplane, then came down hard into the trough of a wave, and all we could see was the crest of the wave entirely around us, the far horizon lost. Once, upon reaching a crest, I looked to my right and saw, no more than a yard away, a Leach's Storm-Petrel, a small gray bird pattering at the water for prey, a bird new to me. What fortune to see one so close.

We caught sight of Great Bird and slowly gained on it, the rock coming closer, the air filling with a confusion of birds, seemingly wingtip-to-wingtip. A snowstorm of birds—just as Audubon had said, when he was here. An immense swarm. Sun streamed down, and, ahead, the heavy, hard surf crashed and shattered against the sides of Bird Rock.

Cartier had seen this wildness, too, when the auks were at Bird Rocks. And after Cartier, the commander of the *Bonaventure* saw Great Auks here in 1591. Three years later another seafarer, Wyet, saw what he called "Penguins." But one Charles Leigh who visited Bird Rocks in the sixteenth century didn't mention seeing *any* Great Auks. It's possible that the species could not breed in years when area fish stocks were depleted, which might explain the absence of auks from Leigh's report.

In 1603, Champlain observed Great Auks in the Gulf of St. Lawrence, calling them "Tangueu." By the time Champlain arrived, the island that stood between Great Bird and the smaller Little Birds or Gull Rocks—about a half-mile away—had eroded. Now marked as a reef on nautical charts, that low shelf was where Cartier had gathered Great Auks. The eroding force of the sea had obliterated one small islet, thus restricting the available habitat to the foot of the cliffs of Great Bird and the Little Birds.

A priest named Father Gabriel Sagard saw Great Auks here between 1623 and 1625. He ate one of the birds, finding it "not inferior to any game we have." According to Canadian writer Farley Mowat, in the late seventeenth century, the traveler Charlevoix "noted that the islets still harboured 'a number of fowl that cannot fly.' "

But by 1700, Great Auks were gone from Bird Rocks. The stories of those who saw and killed them vanished, too. We have only a handful of subsequent brief mentions of the Great Auks of this place. In 1873, C. J. Maynard—he of the failed attempt to shoot Carolina Parakeets hidden in a forest—came to Bird Rocks to collect specimens and write up accounts of bird behavior. He risked death by dangling from a rope near the menacing bills of Northern Gannets and other birds nesting on the cliff ledges. Maynard also came to the Magdalen Islands to learn what he could of the Great Auk. "I made inquiries of every one I met who was interested in birds, about the penguin, for so the great auk was commonly called, but received nothing satisfactory . . ." (That is, until he received, too credulously, an inaccurate report about the bird's persistence in more northerly climes.) By the time permanent settlers arrived in this archipelago, the Great Auk was decades-gone.

Here in a place where the low arctic and boreal ocean zones mingle, I felt in-between past and present. A mixing zone. I listened to the croaks of swarms of Northern Gannets, closed my eyes and heard instead the croaks of Great Auks. If I tried hard enough, I told myself, maybe I could imagine the soft sigh they exhaled, with wings extended, when they died. I opened my eyes.

Bewildered, I watched the life before me, thousands of birds. Suddenly, a Thick-billed Murre—a species I had been trying to spot—flew only two yards overhead. I could see the thin white line along the base of its bill, a mark to distinguish it from the similar Common Murre. At a distance, they look the same: chunky black-and-white alcids. I slapped the rubber side of the boat and exclaimed, "Thick-billed Murre, about damn time!"

As Gaston pulled us to within a few tens of yards from the threatening wave-crash against the rocky shoals and base of Bird Rock, I watched silently the awesome, streaming violence of this place of rock, water, wind and sun. Pascal lifted his camera, tried to focus, then laughed and shook his head at the difficulty of snapping a picture in such pitching surf. Gaston and Nadine watched for shoals. Then Elizabeth said, "Look!" and, in front of us, like little wet monks emerging from a dive, there appeared the heads of seals, their eyes watching a crazy craft of humans in the heavy swells.

The boat heaved up and down as Gaston slowly maneuvered us around the northwest side of the cliff, sheer, shelved, huge, cragged, cracked and cloaked with birds so thickly it was mottled, cobbled. Guano streaked down like icicles. We soon edged along the brightly sunlit eastern side of Great Bird. I struggled to take a few pictures.

I memorized the sight of horizontal ledges lined with birds, one ledge just feet or yards above another, and the amazement of the moment held back the queasiness. Though I knew the low shelves in and near the water would not have been here 300, 400, 500 years ago, I saw in my mind the Great Auks surfing in, hopping upright and waddling away.

I spied two other alcids in addition to the murres: orange-billed Atlantic Puffins and a smaller version of the Great Auk, the Razorbill. Puffins nest in burrows; their white eggs become stained red with the island's sandstone dirt. On one ledge, Razorbills mingled with puffins, all content to stay put, unlike the gannets, who leapt ceaselessly into flight, whirled in wide circles, landed, flew again. I craned my neck to look up the cliff. The wind, hard and ceaseless, kept the smell of guano from staying near one's nostrils for too long. What unmitigated forces! The sheer violence of the seas, the winds. *This is the wildest place I have ever been*, I whispered, then turned to look at Elizabeth, wanting to yell, to proclaim our happy insignificance.

Though I caught glimpses of the lighthouse and other buildings atop Bird Rock, the usual absence here of human lives just made stronger the factual surge of wildness. The ceaseless surge of time, water, wind, this billowing spumescent, vertiginous fracturing whole, white, eerie green, cobalt, red-cliffed, ledged rock that sometimes falls and crashes, and everywhere this force, and feathers white as wave-crash.

I breathed in deep, and we rounded the southwest tip of Bird Rock, which is lower there than the other parts of the cliff. Gaston backed us away from the currents that swirl around a rock arch that protrudes from the cliff like a finger dipped in the water. Along the top of that finger and back on the cliff there were neat lines of gannets.

Had we been here in the prior centuries of the Basque, Breton, English and other fishermen, we would have seen walruses. But they were hunted out like the Great Auks. On the Magdalen Islands today, an elderly but spry man named Leonard Clark studies the histories of walrus kills and shipwrecks. On one night in 1765, more than 800 walruses were killed on the shore near where his house now stands. The first walrus hunt was in 1591, overlapping for a time with the killing of auks. The last walrus hunt was in 1777. "Here look," he told me one evening. "That killed a walrus." He placed in my hand one, then three more, black metal balls. "They were fired point-blank." One ball was flattened from having gone through the head of a walrus or "sea cow."

With no walruses left to prey on seals, and an end to widescale human hunting of them, the seal population has increased, making that species a convenient target on which to place blame for the current collapse of the Gulf fisheries. (Fishermen harbor animosity against fish-eating birds, too.) Doubtless, the existence of more seals has meant more seals eating fish. But this point misses the larger context. The Gulf of St. Lawrence has become the scene of a vicious cycle of capitalist competition and survival as small-scale fishermen try to hold their own against huge fishing vessels—more like factories, really—that send out long, hook-laden lines and bottom-scraping nets. A government-imposed cod and perch moratorium may yet lead to a rebound in the fisheries, some say, but, for now, catch numbers have declined while values have increased. Fewer people are in the business but they can earn higher prices. When Pascal was young a snow crab brought 50 cents; now the price is $7 or $8. Some fishermen have formed cooperatives to share profits and cut losses as the cod fishery reopens. Others are turning to commercial farming of blue mussels.

As Gaston held the boat more or less in front of the western side of the cliff, where a tiny wharf and broken ladder remain, disintegrating, I thought of all the naturalists who had come to this wild place, after the disappearance of the Great Auks from these waters. N. S. Goss of Topeka had come here. Alfred Gross had come here. Ornithologist William Brewster visited in the 1880s and, in addition to writing a delightful account of a pet Black-legged Kittiwake that had to be shown how to land, recorded a terrible gale, which hammered his ship while birds flew calmly above. Naturalist and photographer Herbert K. Job came twice to Bird Rock in the early 1900s. On one of his arrivals, the lobster boat he was in "hurled . . . against the pile" at the cliff's base. And that was on a fairly *calm* day. Nonetheless, Job pluckily concluded, "It takes a windy day to show Bird Rock at its best."

Job had a keen appreciation for the wonders and dangers here. "Reach it in an open boat and find it impossible to land," he wrote, "and there is no telling what might occur. Let the wind breeze up strong, contrary for a return to shelter, and where would one be? Let the ever-ready fog shut down, and what assurance is there of finding that atom [of a boat] amid the great waters, or even of returning to the Magdalen Islands?" After one of his successful journeys to and from Bird Rocks, Job received an honorary degree, given to him by a Madelinot fisherman, for having gone there in an open boat: the *S.S.D.* The Sad Sea Dog.

Even a partial list of shipwrecks at Rochers aux Oiseaux underscores the hazards of the place. *The Nimble*, 1822. *Good Intent*, 1856. *Annie Flemming*, 1877. *Heidi*, 1881. *President Sverdrup*, 1881. *Imperial*, 1881. *S.S. Thanemore*, 1881 (a bad year, evidently). *Maxwell Pellock*, 1890. *Dominion*, 1903. *Loredore*, 1956. *Lady Audette II*, 1971. More than once, crews and passengers—usually European immigrants—crashed at Bird Rock and survivors climbed aboard the nearby slope of Gull Rocks, without food or fresh water, waiting for days to be rescued.

In 1828, a British captain complained about the lack of lighthouses in the dangerous waters surrounding the Magdalen Islands archipelago, but the first lighthouse in the area, the one here at Great Bird, didn't begin operation until 1870.

A person inclined to believe in ghosts might well suspect the Apponatz and sea cows left in their extirpations a curse on the keepers of the light. An 1897 letter from a Canadian fisheries official to an administrator in the Crown Lands Department in Toronto spelled out the first few disasters that befell the lighthouse keepers and their families on this remote station:

> ... Bird Rocks has had a very bad record. The first keeper went out of his mind and with his assistant, was removed in consequence. The second

keeper with his son and assistant left the Island to hunt seals on the ice, and the wind sprung up; they were carried away and never heard of again, excepting the assistant who managed to reach shore nearly frozen to death. The third keeper and his son were killed by the explosion of a keg of gunpowder near the fog cannon [which was fired during heavy cover to warn ships of the location of the rock]. The fourth keeper was nearly killed by the premature discharge of the fog gun. His nephew and two other men died from exposure while seal hunting last winter.

Men adrift on ice drank seal blood and ate raw flesh in order to survive unbelievably cold winters. Utter isolation for months. Gales. The keepers and families at Bird Rock endured that and more.

They pleaded to uncaring administrators about the need to be compensated for using valuable supplies to care for shipwreck survivors. They sometimes had to signal passing ships just to get drinking water. In 1922, men were poisoned by tainted drinking water in catch basins atop the rock. In 1928, a keeper complained, "In the winter, we are forced to gather up snow on the Island which is very dirty and even when the spring rains arrive, we get very little water . . ." Two years later, an assistant wrote of how he, his crew, his wife and child had been abandoned with insufficient supplies by the drunken head lighthouse keeper. Lightning set buildings ablaze in 1955. For decades, resupply and relief remained uncertain, given the tricky weather of the Gulf.

Despite Bird Rocks having been declared a bird sanctuary in 1919 (or 1921, depending on the source), shooting and egging of many species continued there well into the twentieth century. This prompted complaints from ornithologists who visited the rocks, often shooting specimens themselves. The lighthouse keepers were in a difficult position, however. If they attempted to enforce the ban on molesting seabirds, they risked alienating sailors whose vessels might be needed to help the keepers in an emergency.

Eventually, with helicopter service starting in 1966, a ban on families at the station, new monthly rotating crew shifts and the extended reach of radio and television signals, life for humans tempered on Great Bird Rock. When I asked Leon Patton—one of the last keepers of the light, who now lives in retirement on Prince Edward Island—if he missed Bird Rock, he smiled and said softly, "Not really. I was ready to get back to civilization, although it wasn't really that bad a place to work." It was work, after all, a job, and so he and the other keepers tended the equipment, played checkers, watched television, observed the birds and put up with noise and stench. Patton's favorite times were on the one or two calm days in a summer month when the keepers could row out in a dory to fish for cod and mackerel. "There were still fish then," he said.

The lighthouse at Bird Rock station became automated in 1988. The only humans allowed to reach the top of Bird Rock now do so by Canadian Coast Guard helicopter; the 152-step ladder is too dangerous. Crews arrive for routine inspection and repair to make sure the powerful lighthouse bulb still displays its signal.

It's as if the stories of this place tell us now: *You can risk coming here, but you must not stay.* We couldn't. Gaston gunned the engine, taking us northwest, away from Great Bird and toward the small islets of Gull Rocks or the Little Birds. (The Francophones call them Rocher aux Margaux.) With my eyes no longer filled by the looming cliff of Bird Rock, I marveled at the water: such a deep blue.

In my wooziness, I consoled myself by recalling how Audubon had felt seasick on his ship, the *Ripley*, when he was in Labrador. When Audubon was at Bird Rock, though, his sickness was worry, concern over the safety of his son out climbing the cliffs for bird specimens and eggs. Audubon came far too late to see Great Auks here, so to obtain a specimen suitable for illustration, he bought a Great Auk skin at a London auction in 1836. He later gave the specimen to his good friend J. P. Giraud, a naturalist and writer, and Giraud then donated it to Vassar College. For a time, Vassar forgot about the specimen, which gathered grime beneath a

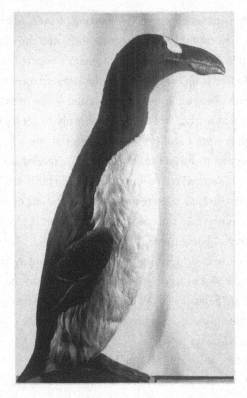

Photo of the Great Auk specimen that, in all likelihood, was the model for Audubon's illustration. The specimen is held at the Royal Ontario Museum.

sink in a classroom lab. The Royal Ontario Museum eventually purchased Vassar's auk, and since 1965 has kept it safely stored in a fireproof vault.

One cold Toronto morning, Brad Millen, of the museum's staff, came in on his day off and opened that vault for me. Not since touching Carolina Parakeets had I been as affected by a bird specimen. Audubon's auk was held within a large glass case, so I could not touch the bird. Brad excused himself while I took notes about the array of specimens before me: Heath Hens, a Labrador Duck, Passenger Pigeons and Ivory-billed Woodpeckers. My eye kept returning to the Great Auk—because, once, *Audubon himself* had held this bird, examined it, had imagined from its shape his illustration of the species.

Now, as we approached Gull Rocks, more tan-colored than the reddish Great Bird, Gaston began to slow in order to keep us clear of the shallow waters around them. The Gull Rocks are three, though two are so close that what divides them is but a narrow crack. From a distance, which was what we kept, one appeared almost like a smaller version of Great Bird, while another seemed to me more like a boulderous column. Herbert Job wrote that the column was about 60 feet high and the smaller version of Great Bird 40 feet high. The water between these sandstone islets is shallow and dangerous. We drifted, and I took my last picture, noting how the rocks had long slopes rising up out of the water—exactly the kind Great Auks would have used. And, invisible, between the bird-festooned Gull Rocks and the huge bulwark of Great Bird lay a reef that once was above water: the "least island" that Cartier had mentioned. Gaston kept us well away, then turned the boat back toward Grosse Île, southwest. I managed one final glance at Bird Rock. It looked less fearsome now, more like a lop-

Audubon's Great Auks.

A view of Bird Rock. Gull Rocks are visible in the distance.

sided cake, one surrounded by a cloud of powdered sugar—innumerable gannets.

We smacked into a vicious head wind, slamming hard against swell after swell, the ride far rougher than before. The water sprayed us as we cut through and on the waves. Elizabeth huddled in her wool sweater and rain gear. She was miserable. I found that putting on my wool hat became a delicate operation requiring many motions: Unzip pocket, rest, reach in pocket, rest, look at what I pulled out, look up, rest. Each movement of my head risked a blackout. I let the horizon—that stable line of sky meeting planet—remind me of a firmer world where I would be no longer dizzy and cold to the bone.

The waters stayed turbulent until we reached the coast of Île Brion, a large uninhabited island between Bird Rocks and the populated Magdalens. Then, as we passed Île Brion, the sea turned nearly mirror-calm, and our moods lightened. Suddenly we could eat and laugh and joke, discussing with amazement the scenes of the day. Gaston gunned the boat, and we moved *fast*. "Six Flags Gaston!" Pascal yelled.

It was late afternoon, and in that late light I watched Common Eiders swimming in the distance. I turned from the others and looked as far as I could. I thought of endings.

After the Great Auks had been exterminated from the Gulf of St. Lawrence, they persisted in North American waters—at Funk Island, off the Newfoundland coast. There, at that huge breeding colony, mariners continued to plunder the birds for food.

At the same time, a traditional, Old World source for mattresses and pillows—eiderdown—was becoming less and less obtainable. The eiders were near extinction. So, from about 1770 to 1780, men came to Funk Island to kill auks solely for their feathers.

The crews built stone corrals into which they herded hapless Great Auks. Each summer, men boiled vats of water and threw the live birds in to loosen the feathers for plucking; that accomplished, the corpses were either thrown to the wayside or used as oily fuel for the fires boiling the water in which more of their kin died. Newfoundland visitor Aaron Thomas wrote in his diary that "while you abide on this island you are in the constant practize of horrid cruelties . . ." He and others like him sometimes would simply take hold of a bird, pluck the feathers, and let it go "to perish at his leasure . . ."

The corpses of auks eventually decomposed into soil. Plants grew up around the bones, and puffins nested amid them. Once, someone even gathered the decayed matter to sell as compost in Boston.

In his journal entry for Tuesday, July 5, 1785, Labrador trapper George Cartwright had warned that this slaughter would doom the species. As well, a missionary on an island close to Funk claimed he often admonished people for killing the birds only for their feathers and he spoke of his dislike for boys who tied the legs of Great Auks in order to keep the birds as pets. Despite a futile enactment of protection for Great Auks as early as 1533, despite Newfoundland's pleading with London in 1775 to ban all exploitations except for food and bait, despite the reported flogging of eggers in 1793, despite the banning by London authorities in 1794 of the killing of Great Auks for feathers, the world's great breeding colony of this species dwindled to none by 1800.

For a time, Great Auks managed to hold on around Iceland. Then in March 1830 volcanic eruptions destroyed the birds' breeding site, an island called Geirfuglasker (Garefowl-skerry). The birds moved to a nearby island, Eldey, where they persisted until June 3, 1844, when two Great Auks still lived on Eldey, two adults tending an egg.

Naturalists John Wolley and Alfred Newton later interviewed many people in order to learn what had happened that day. The morning of June 3, a 14-member fishing crew, hired by an Icelandic collector named Carl Siemsen and captained by a man named Vilhjálmur Hákonarsson, arrived at a low-lying ledge on Eldey. Newton described what happened next when Jón Brandsson, Sigurdr Islefsson and Ketil Ketilsson saw the two Great Auks amid a colony of Common Guillemots:

> The Gare-fowls showed not the slightest disposition to repel the invaders, but immediately ran along under the high cliff, their heads erect, their little wings somewhat extended. They uttered no cry of alarm, and moved, with their short steps, about as quickly as a man could walk. Jón with outstretched arms drove one into a corner, where he soon had it fast. Sigurdr and Ketil pursued the second, and the former seized it close to the edge of the rock, here risen to a precipice some fathoms high, the water being directly below it. Ketil then returned to the sloping shelf whence the birds had started, and saw an egg lying on the lava slab, which he knew to be a Gare-fowl's. He took it up, but finding it was broken, put it down again . . . All this took place in much less time than it takes to tell it. They hurried down again, for the wind was rising. The birds were strangled and cast into the boat . . .

Unfortunately for Carl Siemsen, who had sponsored this expedition, the crew met another man, Christian Hansen, who bought the last Great Auks known to have been alive. For a few years, some observers reported seeing lone Garefowl, but none of the reports could be confirmed.

The body parts of the Eldey auks are today kept in specimen jars at the Universitetets Zoologiske Museum in Copenhagen. No one knows precisely what happened to the skins. Naturalist Errol Fuller, in his recent account of the species, suggests that one of the Eldey auks may be a specimen in the Natural History Museum of Los Angeles County. Another may be a specimen at the Royal Institute of Natural Sciences in Brussels.

Even before the Great Auks had been killed off, a byzantine culture of collectors developed, attracted to the strangeness of the bird and its disappearance. Specimens and eggs, real and fake, became sought-after items. By the end of the nineteenth century in England, the price of a Great Auk egg ran 2 to 4 times—even 11 times—higher than the annual wage of a skilled worker.

Eventually, the bird would come to decorate a cigarette tin in England—Great Auk Cigarettes—as well as the cover of the official journal of the American Ornithologists' Union.

Throughout my time on the islands, I remembered the story of June 3, 1844. And I thought of John Wolley and Alfred Newton investigating, in the 1850s, the life and death of the Great Auk, then so recently gone. To them, as me, the stories mattered. I felt fortunate to have seen at least one place where this species had lived, a place where Great Auks had come ashore to breed. But I felt a keen sense of failure that I had not unearthed anything new about the penguins of Rochers aux Oiseaux.

While I rode the boat back to Grosse-Île, images of Bird Rocks mingled together with imagined scenes from Eldey, from Funk Island, from the Grand Banks. Of Wolley and Newton's pursuits, Newton wrote, "Our science demands something else—that we shall transmit to posterity a less perishable inheritance." I felt the burn of wind on skin, looked at the wide Gulf around me and recalled: While in Iceland, Wolley and Newton noticed something odd in a churchyard wall. In the turf that had been used as a mortar between the stonework, something stuck out. They looked more closely. It was the bone of a Great Auk.

I took comfort from the fact that now Bird Rocks and Funk Island are protected seabird-nesting colonies. But birds there still face pernicious threats: oil pollution, ensnaring trash and nets that tangle birds, poaching and the insanely short-sighted management of the Atlantic fisheries. Such mismanagement not only forced fishermen to knowingly harvest at unsustainable levels, it has put all species dependent on that marine ecosystem at grave risk. Though the Atlantic Puffin is being reintroduced to formerly occupied rocky breeding islands, its population is declining. So, too, the Common Murre and Razorbill. The Thick-billed Murre, apparently, is doing better. Northern Gannets are increasing in number after a nearly disastrous decline in the late nineteenth and early twentieth centuries, but Leach's Storm-Petrels are decreasing. All these birds need the kind of regard and care being given species like the Piping Plover, which nests on beaches of the Îles de la Madeleine, under the vigilant eyes of a local environmental group, Attention FragÎles.

As we neared Grosse-Île, Nadine looked at me till I looked back.

"They must have been beautiful birds," she said.

I said, "Yes, they were," and almost added, "They all are."

On one of our last nights, I walked out on the front porch of our chalet, which overlooked a field of yet-to-bloom wild strawberries and the Lagune du Havre aux Maisons, where, each evening, many Great Blue Herons gathered to stand in the mirroring water. The wind chilled me, but not unpleasantly, and the sky was clear, dark and starry. Elizabeth nestled into a sleeping bag on the porch, and I walked around back and saw the steep brawn of Mont Alice, a nearby hill that loomed black against the stars.

With the chalet behind me, I was out of the glare from a streetlight on the gravel road. I closed my eyes to let them adjust to the dark. I listened to the wind. Then I looked, to the southwest, over the tops of charming cottages, and stared straight into the fire heart of the galaxy, 33,000 light years away. That far, yet here was its cloudy light, spilling all about like sea fog, like spume and spray among the stars of Sagittarius and Scutum. From memory, I looked from star to star, from one globular cluster to another, then on to greenish smears of nebulae, the Trifid, the Lagoon and there, above sharp clusters of stars, the Swan. In these nebulae, so reliable to watch because they appear again each summer, vast clouds of gas and dust congeal into thousands and thousands of suns.

High overhead, birds flew in this dark. Their kind I could not name. Wind sheared their calls into something fantastic and changed. In seasons of passage, in spring and fall, birds use stars to navigate their way, from one distant place to another. Their bodies, their brains, have come to believe in things that abide, things that can show us the way from here to there. I thought of *The English Pilot*, which noted flocks of Garefowl on the Grand Banks: living markers for the weary mariners to know that they had reached the coast of their dreaming.

The stars recede from us, even as we try to reach them. We cannot sully them all. They are, I think, more beautiful from this ground, even if, within the span of my life, up to one-fifth of the species on this ground, this planet, will likely be gone. It's as if we live in a house from which, each hour, we remove a foundation stone, a joist, a rafter. The house could fall to wreckage surrounded by weeds. I do not wish to grow old on a planet of weeds.

On good days, I believe in our capacity to heal, to form communities of concern and action, to recognize loveliness and protect it. On good days, I smile to hear the caroling of a Carolina Wren and know that we are capable of adoration. On bad days, I have to find comfort in the sweep of cosmic and geologic time, wait till night and watch the stars.

If we are to save what is left and restore what we can, then those of us who love birds must do uncomfortable things. Write to politicians. Write them again when their answers do not satisfy. Speak up at county and city commission meetings in defense of the green world. Take local leaders on bird-banding trips so they can stand gape-jawed at the beauty of a Cape May Warbler or that little feathered nova, the Common Yellowthroat. We ought not underestimate the elegance of individual decisions coupled with communal actions—a bird seen, a refuge protected, a vote changed—especially as they accumulate, one by one, the way the barbs and barbules of a feather hold together.

Once, I attended a meeting in Great Bend, Kansas, where a group of state wildlife officials, hunters, biologists, birders and farmers had gathered to discuss the relative merits of expanding the area in which Sandhill Crane hunting was to be allowed. Toward the end of the meeting, one of the farmers bent forward at the table and tilted his head, about to speak. He had been to that point absolutely silent. My stomach tensed. *Here we go*, I thought, *here comes rage*. I expected a fury of outlandish claims. After all, there are some in my state who believe, nonsensically, that the government secretly stocks mountain lions rather than sell more deer-hunting permits. But the man spoke with dignity and grace. He'd been farming on his own for years, he said. But the birds were going away. With the hint of an accent native to those fields, he said, "The air is getting empty."

I remembered the first time I'd ever visited the wildlife refuge at Cheyenne Bottoms, just outside Great Bend. The Bottoms are a large complex of globally important wetlands, home to many creatures and a resting place for many of the shorebirds that migrate through the central United States. Within minutes after pulling onto a gravel road atop a low dike, I saw a Northern Harrier appear just beyond the car's open windows, floating at eye level, patrolling the marsh's reedy perimeter on his out-spread wings. The harrier sailed slowly by, his head turning, looking for prey and looking, for a moment, at me. Cheyenne Bottoms is threatened now by a proposal to build a hog slaughterhouse in its watershed, but local farmers and out-of-town environmentalists have joined to fight the plan.

When I began this book, I wrote that we must redefine hope from wish to work, or, as Bill McKibben says, "Real hope implies real willing-ness to change . . ." What we have lost—and what we have now—oblige us to savor and save what we can. My wife once told me that dedication can be married to delight, and I believe this is so. Often I have attended all-day gatherings of Audubon Society volunteers at which we take hours to discuss the fate of places like Cheyenne Bottoms or analyze the latest anticonservation measures in the legislature or, sometimes, relish a suc-cess—perhaps a local group has acquired a woodland preserve. Then, in the near-dusk light, we have stepped into tallgrass prairie or stood by a wind-ruffled lake and reveled in the honeyed light, in the sanity of friend-ship. Standing with friends, with these colleagues who teach me the names of flowers and tell me stories of birds, I have watched pelicans fly-ing near, close enough that their wing beats sent breezes across the skin. The astronomer Harlow Shapley once wrote, "To be a participant is in it-self a glory."

Of course, sometimes right action—civic activism—seems to drain away not only hours of the day but part of the spirit. Where does atten-tion turn, after a lost battle, after another small wounding we failed to heal? Sometimes, the mind turns inward to examine how we live and what we live for. In such a test one can fall short. My life is buoyed by a superstructure of engines. I have not disowned that. I have not given up the accoutrements of middle-class American life, such as summer travel to mountains, where my wife and I love the tonic of muscle aches, our backpacks leaned against rock. Like many, we perform the rote chores of the perpetually guilt-ridden. We recycle. We pay a little extra to have a portion of our electricity generated by wind power instead of coal hauled to Kansas from Gillette, Wyoming. We belong to a local organic foods cooperative. We are keenly aware that our ability to do much of this is a result of a privileged economic status that many—most—do not have.

And, to be quite honest, we are frightened enough of the future and certain enough of the toll 6 billion humans are taking on the planet that we have decided not to have children. How I have hesitated to say that, because I say it with resignation, not righteousness. Others reach a different conclusion. Our decision makes us, sometimes, quite sad. But I cannot imagine explaining to a child of mine what the Loggerhead Shrike *used* to be, as it perched on a locust tree beside a Nebraska marsh, or what the song of the Hermit Thrush *once* sounded like, echoing among limestone hills and ferns in an Indiana forest.

On the islands, in a northern gulf, standing beneath the stars, I sought out one among many, Sabik, a pale yellow star whose name means *the preceding*. I thought of the sentences I had written and those yet to be. I thought of how humbled I had become in the face of this daunting and dangerous history of vanished—of vanquished—birds. I have learned much from this history and have realized, finally, that sadness at loss is our best *first* response. It should not be our only response. We know the world gives us life, beauty and solace. We would be ungrateful if we failed to give that back.

Afterword

NEAR THE END OF EACH DECEMBER, I PULL OUT MY PLANNER FOR THE
coming year and pencil in important reminders—upcoming meetings at
work, conferences, birthdays, the date of my mother's death in 2001. The
mundane and the intensely personal. And on each February 21, I write
"Last CP Died." Of all the anniversaries of bird extinctions, this date
above all others calls to me over the years. I suppose it's because this
book owes its beginnings to the day I saw Black-hooded Conures in
Kansas, which led me soon enough to the stories of the Carolina Parakeet.
To this day, I think about those improbably bright and vivacious birds; and
even here, in my new home in northern Utah—not so new, I suppose,
having lived here for a few years now—I imagine them against a lattice
of cottonwood limbs and the diamond-hardness of a late summer sky. Of
course the parakeets were never here in Utah, but I expand their range
with imagination and memory. I still miss them. I miss them all.

It would seem that many of us do. Since the first publication of *Hope
Is the Thing with Feathers*, activists, birders, school children, college students and even a governor have written to share with me their thoughts
about these birds and how their stories can inspire us to action. Perhaps
the most striking response came from a woman who had the image of a
Carolina Parakeet tattooed on her back.

Since the original hardback and paperback editions of *Hope Is the
Thing with Feathers*, others have written books, scientific studies and

many popular articles concerning the species I profile. I've also learned of additional material that readers may find of interest.

In 2004, the first book devoted to the Carolina Parakeet was published. Noel F. R. Snyder's *The Carolina Parakeet: Glimpses of a Vanished Bird*, from Princeton University Press, covers the natural history of the species and asserts that the birds lasted much longer than most (including myself) believe. As a compilation and synthesis of material on the Carolina Parakeet, including the pioneering work of Daniel McKinley, the book is invaluable. Readers can judge for themselves what to make of Snyder's anecdotal accounts of the bird surviving into the 1930s and perhaps as late as the 1940s. The stories certainly include some colorful characters. Snyder assigns disease factors greater prominence in the extinction of the species than the effects of hunting, habitat loss and the European honeybee. Here too readers can decide for themselves. I still hope that McKinley's many articles on the Carolina Parakeet might be collected and published in book form. Those articles and Snyder's slender but landmark book represent the two best scientific sources for information on this species.

Glen Chilton has completed a book on the Labrador Duck, *The Curse of the Labrador Duck*, to be published by HarperCollins Canada and Simon & Schuster. Chilton traveled the world in an attempt to examine every stuffed specimen of the Labrador Duck. As well, he and coworker Michael Sorenson published a piece in the July 2007 issue of *The Auk* concerning genetic evaluations of nine eggs said to be from Labrador Ducks. Sadly, they found that none of these eggs came from this species, though the study broke new ground as a pioneering effort to retrieve and study avian DNA from old eggs.

It should be noted once more that Errol Fuller's *The Great Auk* was published by Abrams in 1999. It is a magisterial account, full of science, story and art. There is likely no other book devoted to an extinct North American bird as comprehensive and as beautiful as this one. A revised edition of Fuller's *Extinct Birds* was published by Comstock, a division of Cornell University Press, in 2001. It is global in its scope.

Scientists including Eric Palkovacs and Jeff Johnson have continued to study the Heath Hen. The 2004 team led by Palkovacs published work in *Molecular Ecology* that found the Heath Hen was genetically different enough from mainland prairie chicken populations to merit consideration as a separate species. This conclusion has been further supported by Johnson with additional work published in *Conservation Genetics* and the *Journal of Heredity*. But because those studies necessarily relied on Martha's Vineyard Heath Hen specimens, the DNA analyses might be skewed—since Heath Hens from the mainland may be more genetically similar to other grouse than those from the island. Unfortunately, mainland Heath Hen samples do not exist in any museum collections that would allow testing this hypothesis.

Despite this shortcoming, the history of the Heath Hen has provided valuable insight into how best to manage other prairie chickens. Johnson has studied the ramifications of the Heath Hen's ecological story for its relevance to managing prairie chickens across North America. And he and his collaborators have a warning: ". . . some current populations of Greater Prairie-chickens are isolated and losing genetic variation. . . . We estimate that these populations will reach the low levels of genetic variation found in Heath Hens within the next 40 years." Scientists, policy makers and conservationists need to ensure that enough habitat remains for grouse populations across North America. They also must consider transplanting birds among various populations to increase genetic diversity.

On Martha's Vineyard, the hopes for bringing the booming of prairie chickens back is very much on hold, even as Nature Conservancy official Tom Chase—and many others who care about the Vineyard—continue to educate and to advocate. Chase tells me that despite the controversy caused by the proposed introduction—or reintroduction, depending on your point of view—"the concept of ecological restoration continues to become more deeply imbedded in the conservation thinking on the Vineyard. In some ways, it may even be surpassing the traditional focus on land acquisition. Most of the local conservation groups are putting effort into removing invasive exotic plants, reintroducing fire on their sandplain properties, or introducing native species. . . . In fact, native plant landscaping for private properties seems on the verge of taking off. . . ." Though the State Forest "still suffers from lack of fire while many acres remain covered by exotic pine plantations," Chase says the Conservancy now has a "professional prescribed fire program which burns land on the Vineyard for its own properties and those belonging to partner conservation groups. We are still a long way from restoring fire at an ecologically meaningful scale, but we're on track, and the day will come." Chase also says the island office of the Conservancy is working on ways to "undevelop" land back to native habitats while helping with the need for affordable housing in one of the most pricey real-estate markets in the United States.

Author Geoffrey Sea, in a book that at this writing is forthcoming *(The Atomic Drive-In Theater: Episodes in Mass Destruction and American Life)*, has uncovered additional new details regarding the shooting of the last known wild Passenger Pigeon, Buttons, or as some call it, the Sargents Pigeon. According to Sea's further work, I apparently erred in the location of the shooting by some distance. Sea's evidence puts the shooting at a different property owned by the Southworth family, one close by to the Sargents Station grain mill on Wakefield Mound Road. Also, Sea now lives in the Barnes home; it was Blanche Barnes who stuffed Press Clay Southworth's pigeon specimen. Her family donated the bird to the Ohio Historical Society. Sea tells me that Blanche Barnes died at twenty-three, likely due to the use of arsenic in her taxidermy hobby. Sea also has found tantalizing material regarding the spiritual role of Passenger Pigeons in various Native American cultures. As well, he accepts the date of the Buttons shooting as March 22, based on curatorial notes with the specimen, although these notes also include a reference to the shooting taking place on March 12. I relied on W. F. Henninger who reported March 24. Sea has issues with Henninger's reported date, but it is still the one I accept. In the end, the date may essentially be unknowable. Sea is working to preserve the area from future nuclear-related projects, in conjunction with the Southern Ohio Neighbors Group and the Sargents Historic Preservation Project.

Certainly the biggest news since *Hope Is the Thing with Feathers* was first published is the purported rediscovery of the Ivory-billed Woodpecker in Arkansas. Two sightings in February 2004 and a video shot on April 25, 2004, led to an announcement in *Science* that the species was not extinct. Tim Gallagher, one of the eyewitnesses to the rediscovery of the Ivory-bill, tells me the view he and another observer had during the second February encounter was "unmistakable." The Ivory-bill "cross[ed] the bayou in front of us at close range...." Another video was shot in the summer of 2004, which Gallagher considers authentic. Emphasizing that there were "at least 15 sightings" of Ivory-bills in 2004 and by "good field biologists," Gallagher is certain the species persists.

I remember that when I saw the first video on the news, I gasped in disbelief then delight.

Gallagher tells the rediscovery story in *The Grail Bird: Hot on the Trail of the Ivory-billed Woodpecker*, a 2005 Houghton Mifflin book. Phillip Hoose's *The Race to Save the Lord God Bird*, a 2004 Farrar, Straus and Giroux book, digs deeper into the history of the Singer Tract and the effort to protect those last birds in the South; Hoose includes chapters on the birds in Cuba and recent sightings in the United States (though prior to the 2004 Arkansas sightings). Other books include Jerome Jackson's 2004 *In Search of the Ivory-billed Woodpecker*, from Smithsonian, which was also issued as a paperback in 2006 by HarperCollins. In 2007, Oxford published Geoffrey Hill's *Ivorybill Hunters: The Search for Proof in a Flooded Wilderness*. Michael K. Steinberg's *Stalking the Ghost Bird: The Elusive Ivory-billed Woodpecker in Louisiana* appeared in 2008 from Louisiana State University Press. And filmmaker George Butler has produced a documentary called *The Lord God Bird*.

Despite the ongoing confidence of a number of ornithologists, including Gallagher and his coworkers, arguments rage over the videos, the sightings and, therefore, over the status of the species. A number of skeptical scientists have challenged the evidence, saying the birds sighted were Pileated Woodpeckers, and, despite further searches, no totally conclusive evidence has surfaced that would silence the critics. Good scientists—experts, mind you—have claimed to have seen the bird since the initial rediscovery. According to Colin Nickerson of *The Boston Globe*, a Cornell Laboratory of Ornithology researcher said she witnessed crows pestering an Ivory-bill in winter 2005. This sighting has been met with doubt in some quarters. Indeed, as the newspaper reported on February 10, 2008, "The rancorous dispute has shaken the usually collegial bird community, with mud-slinging between prominent biologists. Doubters last year used a professional journal to accuse the ivory-billed scientists of practicing 'faith-based ornithology.'" Searching continues as of 2008, with scientists using everything from waders and satellites to helicopters and automatic cameras.

I fear that I'm no longer as sure as I was that the bird persists. We may have more books on the Ivory-bill than Ivory-bills themselves. But the sightings have had the salutary effect of aiding the Nature Conservancy's important efforts to acquire thousands of acres of bottomland forests in the American south. The story continues to unfold, and Tim Gallagher tells me to remain hopeful. I will—even as I keep my attention on those conservation matters I can actually help effect here in Utah, from open-space protection in my home of Cache Valley to the crucial work of a group such as HawkWatch International on whose board I now serve.

In the years since this book was first published, the torrent of bad news has sometimes felt overwhelming. Extinctions continue. The world warms. But in the midst of this, it is now apparent that nations around the planet understand that we must meet the challenge of a changing climate and all the consequences—from erratic weather to altered habitats—that such a future poses. Scientists are beginning to see complex shifts in bird migration patterns as responses to the new climate. For example, several Mexican species are seemingly moving into the United States. If their populations can be sustained by appropriate habitat, this will be a boon to birders seeking new species for their life lists. But it could represent the demise of some previously established species in the United States. Of course, it's incumbent on birders to help support habitat-protection measures not only in the United States, but in Mexico and other nations. It's also important to note that a great deal of work on behalf of birds and habitats rests on the volunteer efforts of birders and bird-banders. In addition to the many things individuals can do to lessen their carbon footprint—from changing transportation habits to supporting wind and solar energy—simply going out and participating in migration counts, Christmas Bird Counts and bird-banding (with appropriate training, of course) can help us understand how birds are adapting to the conditions at present.

Depending on your point of view, the predicted rise of cloned animals from presently extinct species might be "fun" or folly. No one has yet cloned any of the birds I chronicle in this book, but efforts continue to clone a mammoth. According to a 2007 article in *The Stanford Daily*, Canadian genetics expert Hendrik Poinar believes the effort can succeed within a couple of decades—in my lifetime, presumably. He acknowledges the thorny ethical morass such a technology might present to a world where present-day habitat preservation and greenhouse mitigation loom as crucial to the survival of humanity and the rest of the natural world.

Artists are doing their part to help us reflect on these matters. For example, Rachel Berwick's powerful installation piece, "A Vanishing; Martha," reproduces the bodies of Passenger Pigeons in amber casts, all set upon a myriad of rods. I saw Berwick's piece in Salt Lake City in 2005, where it was shown as part of the University of Utah's Symposium on Science and Literature. I was moved to tears. Cornell University sculptor

Todd McGrain is launching a series of memorials he calls the Lost Bird Project, with the hope that sculptures of extinct birds can be placed in historically meaningful locales.

Meanwhile, conservation groups continue to work hard against mindless greed and overdevelopment. For those dedicated to the cause of protecting birds and their habitats, the National Audubon Society and the American Bird Conservancy are at the forefront. State ornithological societies have the advantage of being more locally focused. Readers can find a wide range of conservation organizations to support with their time, money and energy, including groups that support renewable energy through carbon offsets. And please don't underestimate the power of educating your political representatives at local, state and national levels. Cynicism never changed a vote, after all. Even in the very conservative state of Utah, good and surprising things are happening. A nearby city council voted against participating in a proposed coal-fired plant. Our current governor has joined a Western-states effort to mitigate the effects of climate change. Utah State University, where I teach, is committed to achieving climate-neutrality.

In early 2007 I traveled to Knoxville, Tennessee, where I gave a reading at the Ijams Nature Center. Finally, I met Nancy Tanner, who herself had seen Ivory-bills in the Singer Tract, with her husband, Jim. It's difficult to express how I felt in meeting Nancy—it was overdue, of course, for we had communicated only by phone and mail. She is a delightful, gracious woman, and welcomed me into her home where she cooked dinner for me and the wonderful staff at Ijams. Her presence at my reading returned to me a depth of feeling I'd had about these birds but had missed for awhile. I had given many presentations from *Hope Is the Thing with Feathers* over the years, perhaps so many that my emotional connection to the material had become somewhat routine. Of course, it's always the current project that really holds a writer's interest, and I had already spent several years working on a book about meteorite hunters. Still. The birds had always mattered, always will. Something happened that night in Knoxville. A flood of feeling and memories returned—scenes from this book were vivid again. A sense of warmth and connection. Recognition of loss. A sense of honor to be reading in Nancy's presence. For the first time in a long while, I choked up as I read the stories of these birds.

While I was staying at Nancy's home, she showed me a photo album from the 1935 Cornell expedition, including photos of the team's drive through Logan Canyon, which opens to Utah's Cache Valley. My home now. If I had known they had taken their photos and recordings of the American Dipper in Logan Canyon, I'd forgotten. Suddenly, knowing that the Cornell team had driven a road in that familiar canyon felt like a long-needed link between my time writing *Hope Is the Thing with Feathers* back in Kansas and my life in northern Utah, where my ecological education continues and where I'm putting down roots.

The night I spent at the house where Nancy and Jim lived for so many years I was awake for a long time. I looked at the dark Tennessee woods, the rain pouring down, and it felt incredibly sweet to think of the Cornell expedition driving through Logan Canyon and Cache Valley. I now often think of this when I see or hear American Dippers in Logan Canyon or even out my back door, where the birds fly up and down the Blacksmith Fork River just paces away from the house.

Though I saw no birds that night in Knoxville, I felt their presence—the owls were out there, awake; other birds were hunkered down for the miserable weather. I watched. I listened to the rain. I felt suffused with the grace and responsibilities of time and place. I felt a poignant connection back to the lives of Jim Tanner and all the others—Alfred Gross and Press Clay Southworth especially—and to the birds they knew, all of them gone, the men and the birds, but still present on these and other pages, alive in words and, so, alive in us.

Acknowledgments

Many people and organizations helped me as I worked on this project. Without the advice and assistance of friends and strangers, I could not have written *Hope Is the Thing with Feathers*. To them I give thanks.

A few individuals assisted me at crucial points in the writing and editing of particular sections. Biologist Daniel McKinley gave my frequent questions much gracious attention, shared portions of his notes and photographs of Doodles, as well as took the time to read my chapters on the Carolina Parakeet. Nancy Tanner, Don Eckelberry and biologists Tommy Michot and Wylie Barrow offered supportive and helpful comments for the chapters on the Ivory-billed Woodpecker. They all shared materials that enriched this book. William Gross and biologist John Toepfer reviewed my chapters on the Heath Hen, providing encouragement and corrections. My friend Dave Rintoul, a biologist and top-notch birder, read material on the Passenger Pigeon and on cloning and did so under extreme deadline pressure. David Ehrlinger gave me access to the Cincinnati Zoo files on Martha and offered useful feedback on my chapter about her. Mary Kruse kindly read material in order to ensure that I had points of her family history correct. Historian James Sherow reviewed passages related to census records and the Flint Hills of Kansas. Biologist Glen Chilton read the chapter on the Labrador Duck and contributed several comments that improved the material. Cloning experts Rob DeSalle and Alan Cooper promptly answered my questions about that field. Author and naturalist Errol Fuller not only sent me copies of John Wolley's original notes on the Great Auk, he encouraged me all along the way and read my chapter on that species.

I also wish to gratefully acknowledge the American Antiquarian Society, which awarded me a 1998 Lila Wallace/Reader's Digest Visiting Fellowship for Historical Research by Creative and Performing Artists and Writers. This residency allowed me to make pivotal progress with this book. I thank the following individuals at the AAS: Ellen S. Dunlap, Nancy Burkett, John B. Hench, Joanne Chaison, Thomas Knoles, Marie Lamoureux, Tim Sheehan, Georgia Barnhill, Russell Martin, Laura Wasowicz, Caroline Sloat, James David Moran, Carol Medico and Bill Young. As well, the summer 1998 fellows helped make my experience productive and enjoyable.

More than once my spirit sagged while thinking about extinct birds. Over the years, a few discussions about this work helped me to continue. For those conversations, I thank Jim Sherow, Mikko Saikku, Dan Flores, Roger Mitchell, Scott Russell

Sanders, Dave Rintoul and Barry Lopez. Some of these people may not have known how much their comments and questions meant to me. They meant a great deal.

During one particularly grim season, when I was volunteering as president of the Kansas Audubon Council and lacked the energy and courage to continue with this project, I happened across a beautiful book called *The Soul of the Night: An Astronomical Pilgrimage*, by the astronomer and writer Chet Raymo. This book, along with the star charts in Terence Dickinson's *Nightwatch*, saved me from bleakness and reminded me of unsullied beauty.

Several other people have read passages or draft chapters and gave me their thoughts: Barry Lopez, Bill McKibben, John Tallmadge, Lisa Knopp, Wayne Dodd, Eric Lagergren, John Price, Angela Hubler, Carolyn Sigler, Jen Johnson, Todd Cokinos, George and Judie Cokinos, Carl Hansen and Elizabeth Dodd. I am grateful to them.

I am responsible for any errors that remain.

To Town Peterson of the University of Kansas and to Brad Millen of the Royal Ontario Museum, I extend my appreciation for access to bird-specimen collections at those institutions.

Don Anderson, Geneva Williams, R. T. Williams, Mike Estes, Richard Pough, Bob Hamilton and Vernon Wright all kindly assisted, in various ways, with my understanding of the Ivory-billed Woodpecker, the Singer Tract and the Tensas River National Wildlife Refuge.

My research on the Heath Hen was enriched by time with Tom Chase, John Varkonda and Robert Culbert. Eulalie M. Regan of the *Vineyard Gazette* copied an enormous number of clippings on very short notice. Susan Burroughs assisted in my work with the Alfred Gross Papers at Bowdoin College. Kristine Hastreiter and the Thornton Burgess Society extended me the courtesy of looking at and quoting from some of his papers. Writer Tom Dunlop generously shared his work regarding the status of forest management on Martha's Vineyard and the "recreation" of Heath Hens. To Jim Carodoza and William Byrne, thank you for showing me those old Heath Hen films.

Jan Dietrich and Edward Maruska, of the Cincinnati Zoo, gave me their assistance as I worked on the story of Martha. George Watson, Phil Angle, John Ruthven and Betsy Nolan-Kunze provided wonderful details about Martha's aviary and her return to Cincinnati. Several people in Pike County helped me in tracking down the story of Buttons: Teddy L. Wheeler, Barb Barker, Lori Burkitt and Blaine Beekman. Curators Carl Albrecht and Robert Glotzhober of the Ohio Historical Society shared files containing information about Buttons. And, of course, to the Kruse family—Mary, Terry, Ted and Vi—I extend warm appreciation.

Amy Wilson, George F. Farr and Art Kieffer of the Chemung County Historical Society in Elmira, New York, assisted me in newspaper research and verifying the location of the Buttonwoods.

The people of the Îles de la Madeleine extended generosity at every turn. I thank Nadine Blacquière and Gaston Arseneau for the ride of a lifetime, as well as Leonard Clark, Mandy Goodwin, Gérald Arseneau, Attention FragÎles, Sébastian Cyr, Lucie d'Amours and Lionel Boudreau. Pascal Arseneau and Céline LeBlanc answered questions cheerfully every day during my time on the islands. Pascal, merci encore pour le dîner.

I survived the completion of the book with the efficient assistance of Kevin Snell and Stormy Kennedy at Claflin Books and Copies. I also wish to thank the Interlibrary Loan Department at Hale Library, Kansas State University, for speedy processing of my myriad, often obscure requests. The staff at KSU's Arts and Sciences Media Center never failed me. Linda Brodersen and Sara Wege, of KSU's English Department support staff, assisted me in a number of chores; for their help I am grateful.

Others assisted in ways too numerous to enumerate, including: George Archibald; Allan Blanchard; Nicole Perron; Denis Chamard; James Throne; Julie Hensley; Mike Hensley; Mike Pelton; Mary Wine; Roy and Maria Barlow; George Chandler; Mrs. Maitland Edey; Alita Prada; Richard Hite; Susan Wilkerson; Bruce Kirby; John D. Stinson; Tony Crawford; Roger Adams; John Johnson; Anne B. Sheperd; Bob Hahn; Doug Hurt; June Starnes; Judy Bolton; Frank Graham; Tom French; Terry Rossignol; Robert Robel; John Zimmerman; Randy Rodgers; Olivia Virgil; Scott Simpson; Peter Van Tassel; Andrew Kratter; Melinda Knapp; Tom Sutherland; Ken Brunson; Mark Lynch; Mary Wine; and Margaret Carroll.

The following institutions and organizations deserve credit for enabling me to conduct necessary research: Cincinnati Historical Society; Ohio Historical Society; New York Public Library; Bowdoin College Library Special Collections; Tensas River National Wildlife Refuge; Louisiana State University Library; Special Collections, Hale Library, Kansas State University; Indiana University Libraries; National Climate Data Center; Midwest Regional Climate Center; Greenville, Mississippi, Public Library; Manhattan, Kansas, Public Library; Dukes County, Massachusetts, Historical Society; Atchison County, Kansas, Historical Society; Cornell University Library of Natural Sounds; Woods Hole Marine Laboratory; Pike County, Ohio, Public Library; and Sous-Centre des Îles Collège de la Gaspesie et des Îles.

Many of my colleagues at Kansas State have provided words of encouragement and a sense of the future, especially my fellow instructor Susan Jackson Rodgers and department head Larry Rodgers. To my creative nonfiction students at KSU, I extend my gratitude. In teaching you, I have learned much. I am lucky to count several of you as friends.

Many people have opened their homes to my wife and me during our long research trips. Thanks to Moira Wedekind and Vladimir Derenchuk, Robert Grindy and Rosemarie King, Roger Mitchell and Dorian Gossy, and Scott Minar and Robin Milliken. An especial thank you to Kathy Conrad-Rutzen and Greg Rutzen. Greg, this book really began with your taking me out to bird on Friendship Road. I'm glad we still walk there.

To everyone with the Kansas Audubon Council, Sustainable Manhattan and the Northern Flint Hills Audubon Society, I offer thanks and support. Friends—far too many to list—you know who you are. Keep up the good work. I'll be rejoining you soon.

I found an important community in the Association for the Study of Literature and Environment. I have been exceedingly fortunate to count a number of the ASLE writers and ecocritics as colleagues and friends, including Cheryl Glotfelty, Scott Slovic, Mike Branch, Allison Wallace, Louise Westling, John Tallmadge, SueEllen Campbell, Ralph Black, Karla Armbruster, Terrell Dixon, David Copland Morris and Suzanne Ross, among many others. To the ASLE nonfiction division that spontaneously formed in Kalamazoo, Michigan, I acknowledge my compatriots John Calderazzo, who gave me sage advice, McKay Jenkins, who told me tales of the book biz, and John Price, who knows his pheasants.

For house-sitting, the new-used computer, friendship and advice, Wendi Tilden and Eric Lagergren have my gratitude through the years.

My family has been supportive and understanding throughout. With love, I thank my mother, Marjorie Cokinos; my sister, Vicki Wright; my niece, the budding poet, Jessica Wright; my father, George Cokinos, and his wife, Judie. I also thank my former in-laws for their support during that time. To my nephew, Nat, please know how much I miss you.

Jeffery Smith and Lisa Werner—those few days in northern New Mexico were magical. Jeff, mentor, had it not been for your proposal, I could not have written mine.

I want to recognize the entire staff at Tarcher/Putnam and Penguin Putnam for their efforts in producing this book. In particular, I thank the helpful, patient and soft-spoken Lily Chin, editorial assistant, and her predecessor, Jocelyn Wright. Meredith Phebus and Alexia Meyers of the Managing Editorial Department devoted themselves to a difficult production schedule and endured my first-time-author questions, as did Judy Margolin of the legal department.

To my agent, Natasha Kern, and her assistants Oriana Green and Laura McDonald, I offer my gratitude for your belief in and guidance of a writer new to the commercial publishing world. I could not have asked for a better agent.

I owe a great debt to my editor at Tarcher/Putnam, the remarkable Wendy Hubbert, who ceaselessly offered up copious amounts of diligence, patience, enthusiasm and editing excellence. The shape of this book owes much to her strong sense of narrative.

Finally, I thank Elizabeth Dodd for the love and support she gave me in the years when we were together.

Selected Bibliography

In the making of this book, I consulted an enormous number of sources, both published and unpublished. For space considerations, I have limited this bibliography to only the most essential and accessible books and articles. Readers desiring the complete bibliography may write me c/o English Department, Kansas State University, Manhattan, Kansas 66506.

Allen, Arthur A. "Hunting with a Microphone the Voices of Vanishing Birds." *National Geographic* 71, 6 (June 1937): 697–723.

Allen, Arthur A., and P. Paul Kellogg. "Recent Observations on the Ivory-billed Woodpecker." *The Auk* 54 (April 1937): 164–184.

Allen, J. A. "The Present Wholesale Destruction of Bird-Life in the United States." *Science-Supplement* 7, 160 (February 26, 1886): 191–195.

American Ornithologists' Union. *Check-list of North American Birds.* 7th ed. Washington, D.C.: AOU, 1998.

Audubon, John James. *The Birds of America.* New York: J. J. Audubon; Philadelphia: J. B. Chevalier, 1840–1844.

Bailey, Harold H. *The Birds of Florida.* Baltimore: The Williams & Wilkins Company, 1925.

Baird, S. F., T. M. Brewer, and R. Ridgway. *A History of North American Birds: Land Birds.* Vol. 2. Boston: Little, Brown, and Company, 1874.

———. *The Water Birds of North America.* Vol. 2. Boston: Little, Brown, and Company, 1884.

Bannon, Henry T. *Stories Old and Often Told, Being Chronicles of Scioto County, Ohio.* Baltimore: Waverly Press, 1927.

Barbour, Thomas. *The Birds of Cuba.* No. 6 of the *Memoirs of the Nuttall Ornithological Club.* Cambridge: Nuttall Ornithological Club, 1923.

Barton, Benjamin Smith. *The Natural History of a Country.* Philadelphia, 1799.

Bartsch, Paul. "A Pet Carolina Paroquet." *The Bird Watcher's Anthology.* Edited by Roger Tory Peterson. New York: Harcourt, Brace and Company, 1957. 148–151.

Bent, Arthur Cleveland. *Life Histories of North American Cuckoos, Goatsuckers, Hummingbirds, and Their Allies.* New York: Dover, 1964.

———. *Life Histories of North American Woodpeckers.* Bloomington and Indianapolis: Indiana Univ. Press, 1992.

Bengston, Sven-Axel. "Breeding Ecology and Extinction of the Great Auk *(Pinguinus impennis)*: Anecdotal Evidence and Conjectures." *The Auk* 101, 1 (January 1984): 1–12.

"Bird Laws." *Science-Supplement* 7, 160 (February 26, 1886): 202–204.

Birkhead, Tim. *Great Auk Islands: A Field Biologist in the Arctic.* London: T & A. D. Poyser, 1993.

Blockstein, David E., and Harrison B. Tordoff. "Gone Forever: A Contemporary Look at the Extinction of the Passenger Pigeon." *American Birds* 39, 5 (Winter 1985): 845–851.

Boardman, Samuel Lane. *The Naturalist of Saint Croix: Memoir of George A. Boardman.* Bangor: n.p. (Privately printed), 1903.

Brewster, William. "Notes on the Birds Observed During a Summer Cruise in the Gulf of St. Lawrence." *Proceedings of the Boston Society of Natural History* 22 (1883): 364–412.

Bridges, William. "The Last of a Species." *Animal Kingdom* 49, 5 (October 18, 1946): 183–186.

Bruette, William. *American Duck, Goose and Brant Shooting.* New York: G. Howard Watt, 1929.

Bryant, Henry. "Remarks on Some of the Birds that Breed in the Gulf of St. Lawrence." *Proceedings of the Boston Society of Natural History* 8 (1861–1862): 65–75.

Butler, Amos W. "Notes on the Range and Habits of the Carolina Parrakeet." *The Auk* 9 (January 1892): 49–56.

Cartier, Jacques. *Navigations to Newe Fraunce.* Trans. John Florio. *March of America Facsimile Series,* no. 10. Ann Arbor: University Microfilms, 1966.

Chapman, Frank M. *Handbook of Birds of Eastern North America.* New York: D. Appleton and Company, 1914.

Chilton, Glen. "*Labrador Duck* (Camptorhynchus labradorius)." No. 307 of *The Birds of North America.* Edited by A. Poole and F. Gill. Philadelphia: The Academy of Natural Sciences; Washington, D.C.: American Ornithologists' Union, 1997.

Christy, Bayard. "The Vanishing Ivory-Bill." *Audubon Magazine* (March–April 1943): 99–102.

Clark, Edward B. *Birds of Lakeside and Prairie.* Chicago: A. W. Mumford, 1901.

Cory, Charles B. *A Naturalist in the Magdalen Islands . . .* Boston, 1878.

Dawson, William Leon. *The Birds of Ohio.* Columbus: The Wheaton Publishing Company, 1908.

Deane, Ruthven. "Some Notes on the Passenger Pigeon *(Ectopistes migratorius)* in Confinement." *The Auk* 13 (July 1896): 234–237.

———. "The Passenger Pigeon *(Ectopistes migratorius)* in Confinement." *The Auk* 25 (April 1908): 181–183.

DeSalle, Rob, and David Lindley. *The Science of Jurassic Park and the Lost World, or How to Build a Dinosaur.* New York: Basic, 1997.

"Destruction of Birds for Millinery Purposes." *Science-Supplement* 7, 160 (February 26, 1886): 196–197.

De Voe, Thomas F. *The Market Assistant.* New York: Hurd and Houghton, 1867.

Doughty, R. W. *Feather Fashions and Bird Preservation.* Berkeley: Univ. of California Press, 1975.

Dutcher, William. "Destruction of Bird-Life in the Vicinity of New York City." *Science-Supplement* 7, 160 (February 26, 1886): 197–199.

———. "The Labrador Duck:—A Revised List of the Extant Specimens in North America, with some Historical Notes." *The Auk* 8, 2 (April 1891): 200–216.

———. "The Labrador Duck—Another Specimen, with Additional Data Respecting Extant Specimens." *The Auk* 11 (January 1894): 4–12.

Eastman, Whitney H. "Hunting for Ivory-bills in the Big Cypress." *The Florida Naturalist* 22 (1949): 79–80.

———. "Ten Year Search for the Ivory-billed Woodpecker." *Atlantic Naturalist* 13 (October–December 1958): 216–228.

Eckelberry, Don. "Search for the Rare Ivorybill." *Discovery: Great Moments in the Lives of Outstanding Naturalists.* Edited by John K. Terres. Philadelphia and New York: J. B. Lippincott Company, 1961. 195–207.

Edey, Maitland A. "The Last Stand of the Heath Hen." *The Dukes County Intelligencer* 39, 4 (May 1998): 155–174.

———. "Once there were billions, now there are none." *Life* 22 (December 1961): 169–176.

Ehrlich, Paul, and Anne Ehrlich. *Extinction: The Causes and Consequences of the Disappearance of Species.* New York: Random House, 1981.

Ehrlich, Paul R., David S. Dobkin, and Darryl Wheye. *Birds in Jeopardy: The Imperiled and Extinct Birds of the United States and Canada, Including Hawaii and Puerto Rico.* Stanford: Stanford Univ. Press, 1992.

———. *The Birder's Handbook: A Field Guide to the Natural History of North American Birds.* New York: Fireside, 1988.

Eifert, Virginia *Men, Birds and Adventure.* New York: Dodd, Mead, & Co., 1962.

Eldredge, Niles. "Life in the Balance." *Natural History,* June 1998, 42–53.

Forbush, Edward Howe. *A History of the Game Birds, Wild-Fowl and Shore Birds of Massachusetts and Adjacent States.* 2nd ed. Boston: Massachusetts State Board of Agriculture, 1916.

———. *Birds of Massachusetts and Other New England States.* Part 2 of *Land Birds from Bob-Whites to Grackles.* Boston: Massachusetts Department of Agriculture, 1927.

Fradette, Pierre. *Les Oiseaux des Îles-de-la-Madeleine: Populations et sites d'observation.* Étang-du-Nord, Quebec: Attention FragÎles, 1992.

French, John C. *The Passenger Pigeon in Pennsylvania.* Altoona: Altoona Tribune Company, 1919.

Fuller, Errol. *Extinct Birds.* New York: Facts on File, 1988.

———. *The Great Auk.* New York: Abrams, 1999.

Gaston, Anthony J., and Ian L. Jones. *The Auks (Alcidae).* Oxford: Oxford Univ. Press, 1998.

Gauthier, Jean, and Yves Aubry, eds. *Les Oiseaux Nicheurs du Quebec.* Montreal: Public en Collaboration avec la Societe quebecoise de protection des oiseaux et le Service Canadian de la faune, 1995.

Gibbons, Felton, and Deborah Strom. *Neighbors to the Birds: A History of Birdwatching in America.* New York: W. W. Norton & Company, 1988.

Gill, Frank B. *Ornithology.* New York: W. H. Freeman, 1990.

Gillmore, Parker. *Prairie and Forest: A Description of the Game of North America, with Personal Adventures in Their Pursuit.* New York: Harper & Brothers, 1874.

Giraud, J. P. *The Birds of Long Island.* New York: Wiley & Putnam, 1844.

Goss, N. S. *History of the Birds of Kansas.* Topeka: Geo. W. Crane & Co., 1891.

Gould, Stephen Jay. *Wonderful Life: The Burgess Shale and the Nature of History.* New York: W. W. Norton & Company, 1990.

Greenway, Jr., James C. *Extinct and Vanishing Birds of the World.* New York: Dover, 1967.

Gregg, W. H. "*Camptolemus labradorius.*" *The American Naturalist* 13 (February 1879): 128–129.

Grieve, Symington. *The Great Auk, or Garefowl* (Alca impennis, Linn.): *Its History, Archaeology, and Remains.* London: Thomas C. Jack, 1885; Edinburgh: Grange Publishing Works, 1885.

———. "Recent Notes on the Great Auk or Garefowl *(Alca impennis).*" *Transactions of the Edinburgh Field Naturalists' and Microscopical Society* Vol. 2. Sessions, 1886–1891. November 23, 1887: 93–119.

Gross, Alfred O. *The Heath Hen. Memoirs of the Boston Society of Natural History* 6, 4 (May 1928). Boston: Boston Society of Natural History, 1928, 483–631.

———. "The Last Heath Hen." *Massachusetts Audubon Bulletin* 15, 5 (1931): 12.

Hahn, Paul. *Where is that Vanished Bird?* Toronto: Royal Ontario Museum and Univ. of Toronto Press, 1963.

Halliday, T. R. "The Extinction of the Passenger Pigeon *Ectopistes migratorius* and its Relevance to Contemporary Conservation." *Bird Conservation* 17 (1980): 157–162.

Hamerstrom, Frances. *Strictly for the Chickens.* Ames: Iowa State Univ. Press, 1980.

Harris, Harry. *Birds of the Kansas City Region.* St. Louis: Transactions, Academy of Science of St. Louis, 1919.

Harwood, Michael. "You Can't Protect What Isn't There." *Audubon* 88, 6 (November 1986): 108–123.

Hasbrouck, Edwin M. "The Carolina Paroquet." *The Auk* 8 (October 1891): 368–379.

Henninger, Rev. W. F. "A Preliminary List of the Birds of Middle Southern Ohio." *The Wilson Bulletin* 9, 3 (September 1902): 77–93.

Herbert, Henry William. *Frank Forester's Field Sports of the United States, and British Provinces, of North America.* New York: Stringer & Townsend, 1852.

"History of the Library of Natural Sounds at the Cornell Lab of Ornithology." <http://birds.tc.cornell.edu/Ins/History/thepast.html>.

Hoage, R. J., ed. *Animal Extinctions: What Everyone Should Know.* Washington: Smithsonian Institution Press, 1985.

Hough, Henry B., ed. *The Heath Hen's Journey to Extinction, 1792–1933.* Edgartown, MA: Dukes County Historical Society.

Jackson, Jerome A. "Ivory-billed Woodpecker." Vol. 5, *Birds, Rare and Endangered Biota of Florida,* edited by James A. Rodgers, Jr., Herbert W. Kale II, and Henry T. Smith, 103–112. Gainesville: Univ. Press of Florida, 1996.

———. "Will-O'-the-Wisp." *The Living Bird Quarterly* Winter 1991: 29–32.

Job, Herbert K. *Among the Water-Fowl.* New York: Doubleday, Page & Co., 1903.

———. *WildWings: Adventures of a Camera-hunter Among the Larger Wild Birds of North America on Sea and Land.* Boston: Houghton Mifflin, 1905.

Johnsgard, Paul A. *Waterfowl of North America.* Bloomington: Indiana Univ. Press, 1975.

Keniston, Allan. "The Last Years of the Heath Hen." *The Dukes County Intelligencer* 7, 4 (May 1966): 282–287.

Kirk, David A. *Status Report on the Great Auk* Pinguinus impennis *in Canada.* Ottawa: Committee on the Status of Endangered Wildlife in Canada, 1994.

———. *Status Report on the Labrador Duck* Camptorhynchus labradorius *in Canada.* Ottawa: Committee on the Status of Endangered Wildlife in Canada, 1994.

———. *Status Report on the Passenger Pigeon* Ectopistes migratorious [sic] *in Canada.* Ottawa: Committee on the Status of Endangered Wildlife in Canada, 1994.

Lammertink, Martjan, and Alberto R. Estrada. "Status of the Ivory-billed Woodpecker *Campephilus principalis* in Cuba: Almost Certainly Extinct." *Bird Conservation International* 5 (1995): 53–59.

Laycock, George. "The Last Parakeet." *Audubon* 71 (March 1969): 21–25.

Le Moine, J. M. *Ornithologie du Canada.* Quebec: Frechette, 1860.

Lewis, E. J. *Hints to Sportsmen, Containing Notes on Shooting; The Habits of the Game Birds and Wild Fowl of America; The Dog, the Gun, the Field, and the Kitchen.* Philadelphia: Lea and Blanchard, 1851.

Lucas, Frederic A. "The Bird Rocks of the Gulf of St. Lawrence in 1887." *The Auk,* April 1888: 129–135.

———. "The Expedition to Funk Island, with Observations Upon the History and Anatomy of the Great Auk." *Annual Report of the Board of Regents of the Smithsonian Institution . . . For the Year Ending June 30, 1888.* Washington, D.C.: Government Printing Office, 1890.

———. "Great Auk Notes." *The Auk,* July 1888: 278–283.

Matthiessen, Peter. *Wildlife in America.* New York: Viking, 1987.

Maynard, C. J. "The Strange and Rare Birds of North America: The Razor-billed Auk." *The American Sportsman,* October 18, 1873: 36.

McKibben, Bill. *Hope, Human and Wild: True Stories of Living Lightly on the Earth.* Boston: Little, Brown and Company, 1995.

McKinley, Daniel. "The Balance of Decimating Factors and Recruitment in Extinction of the Carolina Parakeet, Part I." *The Indiana Audubon Quarterly* 58, 1 (February 1980): 8–18. (May 1980): 50–61. (August 1980): 103–114.

———. *The Carolina Parakeet in Florida.* Gainesville: Florida Ornithological Society Special Publication no. 2, 1985.

———. "The Carolina Parakeet in Pioneer Missouri." *The Wilson Bulletin* 72, 3 (September 1960): 274–287.

Mershon, W. B. *The Passenger Pigeon.* New York: Outing, 1907.

Millen, Brad. "The Great Auk: Our Specimen: Is it Audubon's?" <http://www.rom.on.ca/ebuff/aukours.html>. <http://www.rom.on.ca/ebuff/auk67.html>.

Montevecchi, William A., and David A. Kirk. "Great Auk *(Pinguinus impennis).*" No. 260 of *The Birds of North America,* edited by A. Poole and F. Gill. Philadelphia: The Academy of Natural Sciences; Washington, D.C.: American Ornithologists' Union, 1997.

Montevecchi, William A., and Leslie M. Tuck. *Newfoundland Birds: Exploitation, Study, Conservation.* Cambridge, MA: Nuttall Ornithological Club, 1987.

Moser, Don. "The Last Ivory-bill." *Life* 7 April 1972: 52–62.

Nettleship, David N., and Tim R. Birkhead. *The Atlantic Alcidae: The Evolution, Distribution and Biology of the Auks Inhabiting the Atlantic Ocean and Adjacent Water Areas.* New York: Academic Press, 1985.

Nevin, David. "The irresistible, elusive allure of the ivorybill." *Smithsonian.* February 1974: 72–81.

Newton, A. "Abstract of Mr. J. Wolley's Researches in Iceland Respecting the Gare-fowl or Great Auk." *The Ibis* 3 (1861): 374–399.

Nowotny, Dr. "The Breeding of the Carolina Paroquet in Captivity." Trans. Paul Bartsch. *The Auk* 15 (January 1898): 28–32.

Nuttall, Thomas. *A Manual of the Ornithology of the United States and Canada: The Land Birds.* 2nd. ed., with additions. Boston: Hilliard, Gray, and Company, 1840.

Pearson, T. Gilbert, ed. *Birds of America.* Garden City, NY: Garden City Books/Doubleday & Company, 1936.

———. "Protection of the Ivory-billed Woodpecker." *Bird-Lore* 34 (July-August 1932): 300–301.

Peattie, Donald Culross. *Green Laurels: The Lives and Achievements of the Great Naturalists.* New York: Simon & Schuster, 1936.

Pettingill, Olin Sewall Jr. *My Way to Ornithology.* Norman: Univ. of Oklahoma Press, 1992.

Phillipp, P. B. "Bird's-nesting in the Magdalen Islands." *Proceedings of the Linnaean Society of New York.* 20–23, 2 (1913): 57–78.

Phillips, John C. *A Natural History of the Ducks.* Boston: Houghton Mifflin, 1926.

Raup, David M. *Extinction: Bad Genes or Bad Luck?* New York: W. W. Norton & Company, 1991.

Raymo, Chet. *Skeptics and True Believers: The Exhilarating Connection Between Science and Religion.* New York: Walker, 1998.

Reports of the [Massachusetts] Commissioners of Fisheries and Game. 1899; 1906–1933, Boston.

Remsburg, George J. (Untitled). *The American Society of Curio Collectors Bulletin* 1, 3 (May 15, 1906): 46.

Ridgway, Robert. *The Birds of North and Middle America.* Part 8. Washington, D.C.: Government Printing Office, 1916.

Roney, H. B. "The Importance of More Effective Legislation for the Protection of Game and Fish." *Chicago Field* 9, 1 (February 16, 1878): 9–11.

———. "Among the Pigeons." *Chicago Field* 10, 22 (January 11, 1879): 345–347.

Sagard, Gabriel. *Sagard's Long Journey to the Country of the Hurons.* Westport, CT: Greenwood Press, 1968.

Saikku, Mikko. "The Extinction of the Carolina Parakeet." *Environmental History Review* 14 (Fall 1990): 1–18.

Sanschagrin, Roland. *Geological Report 106 Magdalen Islands.* Quebec: Province of Quebec, 1964.

Schaff, Morris. *Etna and Kirkersville.* New York: Houghton Mifflin, 1905.

Schorger, A. W. *The Passenger Pigeon: Its Natural History and Extinction.* Madison: Univ. of Wisconsin Press, 1955.

Schroeder, M. A., and L. A. Robb. "Greater Prairie-Chicken *(Tympanuchus cupido)."* No. 36 of *The Birds of North America,* eds. A. Poole and F. Gill. Philadelphia: The Academy of Natural Sciences; Washington, D.C.: American Ornithologists' Union, 1993.

Sennett, George B. "Destruction of the Eggs of Birds for Food." *Science-Supplement* 7, 160 (February 26, 1886): 199–201.

Shipley, John R. "The Story of the Chicago Mill and Lumber Company." Paper presented to the Washington County (MS) Historical Society, September 28, 1980.

Short, Lester L. *Woodpeckers of the World.* Monograph Series Number 4. Greenville: Delaware Museum of Natural History, 1982.

Shufeldt, R. W. "Anatomical and Other Notes on the Passenger Pigeon *(Ectopistes Migratorius)* Lately Living in the Cincinnati Zoological Gardens." *The Auk* 32 (January 1915): 28–41.

———. "Death of the Last of the Wild Pigeons." *Scientific American Supplement* no. 2024 (October 17, 1914): 253.

———. "Personal Recollections of Extinct and Nearly Extinct Birds." *The Conservationist* 3, 5 (May 1920): 74–76.

Steadman, David. "Human-Caused Extinction of Birds." In *Biodiversity II: Understanding and Protecting Our Biological Resources.* Edited by Reaka-Kudla, Marjorie L., Don E. Wilson, and Edward O. Wilson. Washington: Joseph Henry Press, 1997.

———. "Prehistoric Extinctions of Pacific Island Birds: Biodiversity Meets Zooarchaeology." *Science* 267 (February 24, 1995): 1123–1131.

Stevenson, Henry M., and Bruce H. Anderson. *The Birdlife of Florida.* Gainesville: Univ. Press of Florida, 1994.

Sutton, George Miksch. *Bird Student: An Autobiography.* Austin: Univ. of Texas Press, 1980.

———. *Birds in the Wilderness: Adventures of an Ornithologist.* New York: Macmillan, 1936.

Tanner, James T. *The Ivory-billed Woodpecker.* New York: National Audubon Society, 1942.

———. "Present Status of the Ivory-billed Woodpecker." *The Wilson Bulletin* 54, 1 (March 1942): 57–58.

———. "Three Years with the Ivory-billed Woodpecker, America's Rarest Bird." *Audubon Magazine,* January–February 1941: 5–14.

Terres, John K. "My Greatest Birding Day." *Bird Watcher's Digest,* July/August 1987: 84–88.

Wharram, S. V. "The Passenger Pigeon in Ohio." *Bird-Life* 39, 2 (August 30, 1943): 65–68.

Widmann, Otto. *A Preliminary Catalog of the Birds of Missouri.* St. Louis: Transactions of the Academy of Science of St. Louis, 1907.

Wilmut, Ian. "Cloning for Medicine." *Scientific American* 279, 6 (December 1998): 58–63.

Wilson, Alexander. *American Ornithology.* Philadelphia: Bradford & Inskeep. 1808–1814.

Wilson, E. O. *The Diversity of Life.* New York: W. W. Norton & Company, 1993.

Woodard, Sarah A. "Early Times in Kansas." *Log Cabin Days.* 24–26. Manhattan, KS: Riley County Historical Society, 1929.

Worster, Donald. *Dust Bowl: The Southern Plains in the 1930s.* Oxford: Oxford Univ. Press, 1982.

Wright, Albert Hazen. "Early Records of the Carolina Parakeet." *The Auk* 29 (July 1912): 343–363.

Wright, Mabel Osgood. *Birdcraft.* 1895. Reprint, New York: Macmillan, 1936.

Zimmerman, John. *The Birds of Konza: The Avian Ecology of the Tallgrass Prairie.* Lawrence: Univ. Press of Kansas, 1993.

Art and Text Credits

240   Courtesy, Mary, Terry and Ted Kruse.
249   Credit: Sara Wege/Chris Cokinos/Sheri Fleener.
253   Photo by Christopher Cokinos.
255   Photo by Christopher Cokinos.
262   Courtesy, the Cincinnati Zoo and Botanical Garden.
268   Courtesy, the Cincinnati Zoo and Botanical Garden.
274   Courtesy, Betsy Nolan-Kunze.
276   Photo by Christopher Cokinos.
279   Courtesy, American Antiquarian Society.
283   Courtesy, American Antiquarian Society.
284   Courtesy, American Antiquarian Society.
293   Courtesy, Hale Library, Kansas State University.
296   Photograph by Christopher Cokinos.
306   Courtesy, Gérald Arseneau.
307   Courtesy, Fisheries and Oceans Canada.
314   Courtesy, American Antiquarian Society.
316   Courtesy, Errol Fuller.
326   Courtesy, the Royal Ontario Museum.
327   Courtesy, American Antiquarian Society.
328   Photograph by Christopher Cokinos.

I gratefully acknowledge the following sources of illustrations: the American Antiquarian Society; the Living History Farm, Urbandale, Iowa; Shirley A. Briggs and Daniel McKinley; the Cincinnati Zoo and Botanical Garden; Sara Wege; Mary, Terry and Ted Kruse; Don Eckelberry; Brad Millen; Betsy Nolan-Kunze; Hale Library, Kansas State University; Royal Ontario Museum; Errol Fuller; Canadian Coast Guard; Gérald Arseneau; Alfred Otto Gross Collection, Special Collections and Archives, Bowdoin College Library; and Tensas River National Wildlife Refuge, U.S. Fish and Wildlife Service, Ivory-billed Woodpecker Records, Louisiana and Lower Mississippi Valley Collections, LSU Libraries, Louisiana State University, Baton Rouge, Louisiana.

I gratefully acknowledge the following sources of copyrighted material who have granted permission for quotation in this book: Permission to quote from Robinson Jeffers's poem, "Bixby's Landing," granted by Jeffers Literary Properties; permission to quote from unpublished work by James Tanner granted by Nancy Tanner; permission to quote from the writing of Press Clay Southworth granted by Mary Kruse; the selection from William Stafford's poem "Our Story" reprinted by permission of the Estate of William Stafford; material from the archives of the National Audubon Society, Papers of John H. Baker, held at the New York Public Library, reproduced by permission of the National Audubon Society; permission to quote from the Alfred Otto Gross Collection, Special Collections and Archives, Bowdoin College Library, granted by Bowdoin College and by William A. O. Gross; permission to quote from Alfred Gross's *The Heath Hen* granted by The Museum of Science, Boston; "Heath Hen Begs Bread While Students Stalk," copyright © *Vineyard Gazette* 1932, reprinted by permission; permission to quote from William C. Mills Correspondence Society Archives (Box 724) granted by the Ohio Historical Society. Permission to quote from the following granted by the Smithsonian Institution Archives: SIA, Record Unit 7252, Edward Alexander Preble Papers, 1887–1957 and undated, Box 8, Folder 22, and Dr. Charles Richmond, U.S.N.M., Memorandum to Registrar, September 4, 1914, SIA, Record Unit 305, Office of the Registrar, Accession Records, 1834–1958, with accretions to 1976, Accession File 57354.

This list excludes those sources that are in the public domain and/or within the limits of fair use.

Index

Page numbers in *italics* indicate illustrations.

About the Author

Christopher Cokinos is also the author of *The Fallen Sky: An Intimate History of Shooting Stars* (Tarcher/Penguin) and the winner of several prizes, including a Whiting Writer's Award, the Glasgow Prize for an Emerging Writer in Nonfiction and the Sigurd Olson Nature Writing Award. He has won fellowships and grants from the National Science Foundation, the American Antiquarian Society and the Utah Arts Council. His nonfiction, reviews and poems have appeared in the *Los Angeles Times, Orion, Shenandoah, Birder's World, Science* and *Poetry*, among many other publications. Cokinos lives with his partner, Kathe Lison, along the Blacksmith Fork River in northern Utah's Cache Valley. He teaches at Utah State University, where he has appointments in English and Natural Resources and where he founded *Isotope: A Journal of Literary Nature and Science Writing*.

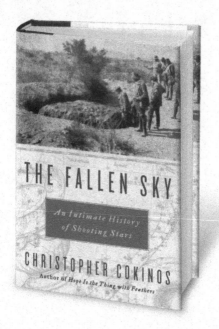